T0130649

Analytical Chemistry
A Practical Approach

Analytical Chemistry

A Practical Approach

E. Hywel Evans and Mike E. Foulkes

E. Hywel Evans
Associate Lecturer at the Open University, and formerly
Reader in Analytical Chemistry, University of Plymouth, UK.

Mike E. Foulkes
Formerly Senior Lecturer in Environmental and
Analytical Chemistry, University of Plymouth, UK.

OXFORD
UNIVERSITY PRESS

OXFORD
UNIVERSITY PRESS

Great Clarendon Street, Oxford, OX2 6DP,
United Kingdom

Oxford University Press is a department of the University of Oxford.
It furthers the University's objective of excellence in research, scholarship,
and education by publishing worldwide. Oxford is a registered trade mark of
Oxford University Press in the UK and in certain other countries

© E. Hywel Evans and Mike E. Foulkes 2019

The moral rights of the authors have been asserted

Impression: 1

All rights reserved. No part of this publication may be reproduced, stored in
a retrieval system, or transmitted, in any form or by any means, without the
prior permission in writing of Oxford University Press, or as expressly permitted
by law, by licence or under terms agreed with the appropriate reprographics
rights organization. Enquiries concerning reproduction outside the scope of the
above should be sent to the Rights Department, Oxford University Press, at the
address above

You must not circulate this work in any other form
and you must impose this same condition on any acquirer

Published in the United States of America by Oxford University Press
198 Madison Avenue, New York, NY 10016, United States of America

British Library Cataloguing in Publication Data

Data available

Library of Congress Control Number: 2018943466

ISBN 978–0–19–965171–9

Printed in Great Britain by
Bell & Bain Ltd., Glasgow

Links to third party websites are provided by Oxford in good faith and
for information only. Oxford disclaims any responsibility for the materials
contained in any third party website referenced in this work.

Every effort has been made to trace and contact copyright holders prior to
publication. Where this has not proved possible, if notified, the publisher will
undertake to rectify any errors or omissions at the earliest opportunity.

This book is dedicated to students and lab-hands everywhere; trying to get the right result.

Preface

Analytical Chemistry: A Practical Approach does not attempt to be a comprehensive treatise on analytical chemistry. Consider it rather as a road manual for the aspiring or practising analytical chemist, either studying the subject for the first time, or requiring an easy-to-follow refresher course or reference book. We have steered away from giving a blow-by-blow account of theory or detailed descriptions of instrumentation, and focused instead on using analytical chemistry to solve analytical problems. This is achieved by taking a problem-based approach, with liberal use of worked examples, student-led activities, and more advanced problems. In this way, we hope that the student can immediately apply what they have learned to similar problems which they must solve at work or in study.

About the book

Analytical Chemistry: A Practical Approach is built around three practical features that will help you to build your understanding of the subject and to apply what you have learned. This section explains how to use these features.

EXAMPLE 1.1

The candy question
The particular candies in question are oblate spheroids wit
(0.2 in) and a major axis of about 15 mm (0.6 in). They con
yellow, green, blue, mauve, pink, and brown, and may each b
shape, mass, and density—a famous brand springs to mind, b
Now we all have our own favourite colour(s), so when we pi
noted in our minds. We may even, after consuming an entir
and obviously biased estimation of the tube contents, for

Examples

Numerous examples provide detailed illustrations of the chapter material, showing you how the ideas from the chapter are applied in practice.

ACTIVITY 1.1 The right questions?

You are a public analyst and a member of the public has just a
they walk into your office, they are carrying a capped plastic b
colourless liquid. Write down what you think are the 10 most
should ask, after they say to you: 'Is this water OK to drink?'

When you have completed this activity, study the feedback gi
chapter.

Activities

Frequent activities give you the opportunity to check your understanding and apply what you have just learned. Feedback on the activities is provided at the end of each chapter.

? PROBLEM 1.1

Definition of terms
You will have come across a great deal of terminology w
as yet not fully defined. Every science has its own jargon
language. This is particularly noticeable when two or m
their particular topic area. With this in mind, you are requ
and phrases in order to see just how important they are t

Problems

Problems at the end of each chapter provide you with additional exercises and extended scenarios that you can use to check your understanding, apply what you have covered in the chapter, and develop your problem-solving skills. Feedback on the problems is provided at the end of each chapter.

To avoid intermediate rounding errors, but to keep track of values and to spot numerical errors, we display intermediate results as n.nnn... and normally round the calculation only at the final step.

About the online resources

The online resources that accompany this book provide a number of useful teaching and learning resources and are free of charge.

The resources can be accessed at:

 www.oup.com/uk/evans_foulkes/

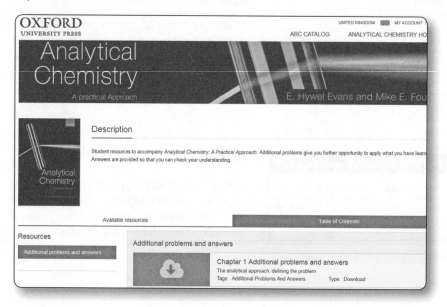

Please note that lecturer resources are available only to registered adopters of the textbook. To register, simply visit **www.oup.com/uk/evans_foulkes/** and follow the appropriate links.

Student resources are openly available to all, without registration.

For students

Additional problems and answers

Additional problems give you further opportunity to apply what you have learned. Answers are provided so that you can check your understanding.

For registered adopters of the book:

Figures and tables from the book

Lecturers can find the artwork and tables from the book online in ready-to-download format. These can be used for lectures without charge (but not for commercial purposes without specific permission).

Acknowledgements

The production of a textbook is never just the work of the credited authors. The content necessarily builds upon and adapts the work of others. Parts are re-written or thrown away after trialling them in lectures and short courses. The activities and problems have been road-tested by undergraduate students over many years. And, not least of all, it incorporates hours spent in the laboratory failing to get the right result. There are too many people to acknowledge individually, but special mention must be made of colleagues and students at the University of Plymouth, without whom this book would not have been written.

Brief contents

Contents

List of abbreviations

AAS	atomic absorption spectroscopy
ac	alternating current
AES	atomic emission spectroscopy
AFS	atomic fluorescence spectrometry
AMS	accelerator mass spectrometry
ANSI	American National Standards Institute
AOI	analyte of interest
APDC	ammonium pyrrolidinedithiocarbamate
ASTM	American Society for Testing and Materials
ASV	anodic stripping voltammetry
BCR	Community Bureau of Reference
CCD	charge coupled device
CE	capillary electrophoresis
CEN	European Committee for Standardization
CRM	certified reference material
CSV	cathodic stripping voltammetry
CV-AAS	cold vapour atomic absorption spectroscopy
CV-AFS	cold vapour atomic fluorescence spectrometry
CZE	capillary zone electrophoresis
dc	direct current
DCM	dichloromethane
DDW	deionized distilled water
DEFRA	Department for Environment, Food, and Rural Affairs
DNA	deoxyribonucleic acid
DSC	differential scanning calorimetry
DTA	differential thermal analysis
EA	Environment Agency
ECD	electron capture detector
ED	energy dispersive (detection system)
EDL	electrodeless discharge lamp
EDTA	ethylenediaminetetraacetic acid
EDXA	energy dispersive X-ray analysis
EDXS	energy dispersive X-ray spectroscopy
Eh	a measure of the redox (oxidation-reduction) state of a solution; the potential of its solutes
EI	electron ionization
EIE	easily ionizable element
EIMS	electron ionization mass spectrometry
ELISA	enzyme-linked immunosorbent assay
EPA	Environmental Protection Agency
ESI	electrospray ionization
ESI-MS	electrospray ionization mass spectrometry
EPR	electron paramagnetic resonance

ESR	electron spin resonance
ETA	electrothermal atomization
ETAAS	electrothermal atomic absorption spectroscopy
ETV	electrothermal vaporization
ETV-AAS	electrothermal vaporization atomic absorption spectroscopy
EU	European Union
FAAS	flame atomic absorption spectroscopy
FAES	flame atomic emission spectroscopy
FAFS	flame atomic fluorescence spectrometry
FFF	field flow fractionation
FI	flow injection
FPD	flame photometric detector
FSA	Food Standards Agency
FT	Fourier transform
FTIR	Fourier transform infrared
FWHM	full width at half maximum
GC	gas chromatography
GC-FPD	gas chromatography flame photometric detector
GC-MS	gas chromatography mass spectrometry
HCCH	Hexachlorocyclohexane
HCL	hollow cathode lamp
HDPE	high density polyethylene
HG	hydride generation
HG-AAS	hydride generation atomic absorption spectroscopy
HG-AFS	hydride generation atomic fluorescence spectroscopy
HG-QT-AAS	hydride generation quartz tube atomic absorption spectroscopy
HMRC	Her Majesty's Revenue and Customs
HPLC	high performance liquid chromatography
HSE	Health and Safety Executive
IC	ion chromatography
ICP	inductively coupled plasma
ICP-AES	inductively coupled plasma atomic emission spectroscopy
ICP-MS	inductively coupled plasma mass spectrometry
ICP-OES	inductively coupled plasma optical emission spectroscopy
IR	infrared
IRMM	Institute for Reference Materials and Measurements
ISE	ion selective electrode
ISO	International Organization for Standardization
IUPAC	International Union of Pure and Applied Chemistry
LC	liquid chromatography
LC-MS	liquid chromatography mass spectrometry
LGC	the national metrology institute of the UK
LLE	liquid-liquid extraction
LOD	limit of detection
LOQ	limit of quantitation
MAC	maximum admissible concentration
MALDI	matrix-assisted laser desorption/ionization
MHRA	Medicines and Healthcare products Regulatory Agency
MIP	microwave induced plasma
MoD	Ministry of Defence
MRC	Medical Research Council
MRI	magnetic resonance imaging

MS	mass spectrometry
m/z	mass to charge ratio
NIES	National Institute for Environmental Studies
NIOSH	National Institute of Occupational Safety and Health
NIST	National Institute of Standards and Technology
NMR	nuclear magnetic resonance
NRCC	National Research Council of Canada
OES	optical emission spectroscopy
PA	public analysts
PAH	polycyclic aromatic hydrocarbons
PBDE	polybrominated diphenyl ethers
PC	polycarbonate
PCB	polychlorinated biphenyl
PET	polyethylene terephthalate
PGM	platinum group metal
PMT	photomultiplier tube
ppb	parts per billion (10^{-9})
ppm	parts per million (10^{-6})
ppq	parts per quadrillion (10^{-15})
ppt	parts per trillion (10^{-12})
PTFE	poly(tetrafluoroethylene)
PVC	polyvinyl chloride
QA	Quality Assurance
QC	quality control
QMS	quality management system
RF	radio frequency
RM	reference material
RNA	ribonucleic acid
RSD	relative standard deviation
SD	standard deviation
SEM	scanning electron microscopy
SI	système international d'unités
SIMS	secondary ion mass spectrometry
SPE	solid phase extraction
SPME	solid phase microextraction
STEM	scanning transmission electron microscope
TEM	transmission electron microscopy
TG	thermogravimetry
TGA	thermogravimetric analysis
TIMS	thermal ionization mass spectrometry
TLC	thin layer chromatography
TS	Trading Standards
US EPA	United States Environmental Protection Agency
UV	ultraviolet
UV-VIS	ultraviolet-visible spectrophotometry
WD	wavelength dispersive (monochromator)
WWT	waste water treatment
XRD	X-ray diffraction
XRF	X-ray fluorescence
XRFS	X-ray fluorescence spectrometry

1

The analytical approach: defining the problem

Learning outcomes

Once you have read through this chapter, worked through the activities and problems, and reviewed the feedback, you should be able to:

1. understand the six steps to solving a problem in analytical chemistry;

2. define a problem in analytical chemistry in a number of scenarios;

3. develop an outline problem-solving strategy applicable to a range of problems in analytical chemistry.

1.1 Introduction

We are all familiar with those dramatized versions of whodunit stories, involving crime-solving laboratories, as presented on television. There is a seemingly effortless flow of information that comes from a trail of events and evidence relating to a crime. In a school, college, or university, where a chemistry course requires a student to perform the **analysis** of a material, it is usually in a controlled environment where the problem is well defined and the procedures are clearly laid out; quite often this is to demonstrate a particular technique or point of chemistry. In real life, this ideal state of ordered affairs is rarely encountered, if at all. Each of the steps involved in **solving a problem** actually requires the asking of questions; the right questions. And while the analytical chemist these days is more likely to be involved with industrial, environmental, clinical and service provider problems, the solution to any problem still starts with and relies on the right questions being asked throughout the process.

The analyst often has to formulate a solution to a problem which comes from two or more apparently different but related questions. Consider the following question: 'Is this bore water sample, taken from my land, fit to drink?' Now compare this with the question: 'Is the nitrate level in this ground water sample above the EC maximum advisable drinking water concentration?' The second question may well relate to the first but at a different (lower) level of complexity. It can be shown that the approach taken to answer both questions and help provide a solution in each case will follow a similar framework. This framework, or **analytical approach**[1] as we shall call it, follows a logical series of steps which can be applied as much to answering the question 'How much antimony is there in mineral waters contained in plastic bottles?', as it can to 'Are the levels of sulfur dioxide being emitted from a gas-fuelled power station above the legal limit?', or even, as we asked before, 'Is this water sample fit to drink?' While the emphasis presented for each of these steps varies from book to book, the general framework can be presented as **six steps** or stages.

1.2 Six steps to a solution

The six steps to solving an analytical problem are:

1. Defining the problem	The 'what', 'why', and 'how'
2. **Sampling**	How you acquire a suitable or **representative sample**
3. Sample preparation	What needs to be performed to get the sample into a suitable form for measurement
4. Measurement	Identifying the most appropriate **measurement technique** to use, the method to apply, and generating the basic data
5. Evaluation	Taking the data from step 4 and turning this into 'values' which can be used for the intended purpose
6. Analytical information	Considering the term **fit for purpose** for the whole process

These steps need to be presented in more detail in order to understand what is going on. In doing so, we are also providing a brief but expanded definition that links with later chapters. Taking these steps one by one we have:

1.2.1 Step 1: defining the problem

The 'what', 'why', and 'how'.

Here, the analyst is really considering a series of important questions, and these may be considered in sequence. The first two questions to consider are: '**what is to be measured?**' and '**why is it being measured?**' From the answers to these two simple questions will come a series of further, more searching questions which will hopefully provide the analyst with the information they require; that is, they will have defined the problem itself, the reason for the analysis, and the property to be measured. The 'how' question is actually addressed to the analyst who considers 'How should I go about tackling this analytical problem and in a safe, **risk assessed** manner?' The answer(s) to this question helps to build up an early framework for all the following steps.

As an example for this first step, let us explore the candy question, that is, 'How much natural food colourant is present in each of eight different candies?' The 'what' question has been partly answered: the **analyte** of interest being the different colourants used for the eight candy colours

---

[1] Some analytical chemistry books refer to the **analytical process**.

and the quantities present for each. The 'why' question will identify the reason and hopefully allow a greater focus on the whole analytical process. To answer this, all food products are monitored and there are regulatory requirements for various components, particularly in confectionary materials, when it comes to additives such as colourants. It will therefore be our task to provide a measure of the colourants present on samples of candies and to compare these against any regulatory requirements. We will explore this interrogative approach in a little more detail later on in this chapter but note that the answer to the 'how' question for the candies example will unfold as we go through the different steps.

Before we move on to the next stage, it is important to note that if this first step is in error and, for example, the scenario is ill defined leading to the wrong analyte being measured, then this step can introduce the largest error in the whole analytical process. If the right questions are not asked and if the answers are not interpreted correctly, then you have the potential of being 100% wrong! While we will be discussing sources of error in the later chapters and evaluating their magnitude within the whole analytical process, these are still based upon the assumption that the problem is correctly defined at the beginning.

1.2.2 Step 2: sampling

How you acquire a suitable or 'representative' sample.

In this step the analyst evaluates the conditions required to take a suitable sample, to help answer the original question(s) posed in step 1. Assuming you have asked all the right questions and defined **both** the problem and **risks (health and safety)** correctly, then this physical sampling step introduces the greatest errors of all the steps, and may require a lot of planning and forethought.

In order to solve any original problem identified from step 1, some form of measurement on a material will be required. To do this, the analyst will need to acquire a suitable representative sample of this material. By representative sample we mean a sample, resulting from a sampling plan, that can be expected to adequately reflect the properties of interest of the parent population.[2] Alternatively, a representative sample can be defined as one that has an identical composition to the larger **bulk** material, batch, or lot, from which it has been taken.[3] As a result, the property you wish to measure on this sample, for example, the density of the sample, presence of a toxic compound, or the concentration of a target element of interest, should be the same as in the original material from which it was taken, within suitably identified limits. This property must therefore also be stable in the acquired sample; it must not change! In technical terms we say that the **sample integrity** must be maintained.

A simple aqueous sample, such as a flavoured still-water drink taken from a large production vat in a beverage factory, where everything is in solution, presents relatively few problems in terms of **sampling errors**. One might consider that all the components in the drink product will demonstrate a **homogeneous distribution**, if the vat is efficiently and continuously stirred, such that a 1 mL aliquot would have the same composition as the contents of the entire 10,000 L vat itself. Unfortunately, this well-behaved example is rarely the case, and acquiring a suitable sample for analysis depends much upon the sample type and its environment. We can illustrate some of the effects associated with sampling by using two examples; the candy question; and the muesli challenge.

---

[2] Horwitz, W. 1990. Nomenclature for sampling in analytical chemistry (Recommendations 1990). *Pure and Applied Chemistry, 62*(6), pp.1193–1208.

[3] Prichard, F.E., Pritchard, E., and Green, J. 2001. *Analytical Measurement Terminology: Handbook of Terms used in Quality Assurance of Analytical Measurement.* Cambridge: Royal Society of Chemistry.

EXAMPLE 1.1

The candy question

The particular candies in question are oblate spheroids with a minor axis of about 5 mm (0.2 in) and a major axis of about 15 mm (0.6 in). They come in eight colours: red, orange, yellow, green, blue, mauve, pink, and brown, and may each be considered to be of equal size, shape, mass, and density—a famous brand springs to mind, but other products are available. Now we all have our own favourite colour(s), so when we pick out our preferred candy it is noted in our minds. We may even, after consuming an entire pack, perform an unscientific and obviously biased estimation of the tube contents, for example, there weren't as many blue ones as usual in this pack! Because all the candies are identical, with the exception of the colour, you have a one in eight chance of picking out your chosen colour. But what is the likelihood of you drawing out all eight colours, in any order you like, from just eight draws, from a bulk of, say 512 candies in a jar, containing 64 of each colour? Now, if each candy weighs on average 1.3 g then what colours do you acquire if you poured out, first 20.8 g and then 52.0 g from your 512 bulk source? And finally, if each candy actually has a mass between 1.2 and 1.4 g, then how many do you actually get in each sample **batch** or **sample increment**, as your **primary sample** is often known, and what colours do they each contain if you weigh out 10 sequential increments each of 52 g from your bulk source of 512? It is worth noting at this point that sampling based on a population (numbers of units) on the one hand, and mass on the other, can provide some interesting differences.

For the purpose of our sampling example using candies and addressing the problem we set ourselves in step 1—how much natural food colourant is present in each of the eight different candies—we will divide our source of 512 into 8 smaller batches, where each batch has 64 of the same colour. From each batch we will draw at random 12 candies and repeat the process four more times so that each single colour batch produces five smaller sub-samples. At the end of the process we have 40 sub-samples with five replicates of each colour. Each replicate is identified by a unique number, ready to be prepared in the next step later on (Figure 1.1).

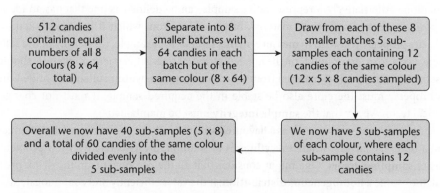

Figure 1.1 Flow chart showing the sampling process for Example 1.1

EXAMPLE 1.2

The muesli challenge

We are enthusiastic eaters of breakfast cereals these days and one of the so-called healthier options available to us is the balanced diet of the muesli mix. The 'fruit and nut' variety comes in a range of recipes including blends of whole rolled oats, wheat and barley flakes, and sunflower seeds with mixed dried fruits such as sultanas and raisins, apricots, dates, and apple, and nuts such as hazelnuts, walnuts, almonds, brazil, and coconut. The label on the packet identifies the percentage by weight of the components together with nutritional information. As any public analyst will inform you, labelling is a very important part of the food markets' presentation,[4] so with labels which must be clear, not misleading, and with a list of ingredients, if two or more are present. There is also a requirement to disclose the quantities of ingredients, in descending order for certain goods. From our above recipe we can have up to 14 or 15 different ingredients or components. Each component is a different size, shape, density, and composition, a very **heterogeneous** mixture, that is, one that varies throughout its extent. Now, given that the 14 or so major components can vary between 25% and 2% by weight and particles may be as large as 20 mm down to 2 or 3 mm with varying shape and density, then the ability to take a **representative sample** becomes quite a challenge. Imagine being tasked with checking the assumed 2% raisin content in a supermarket brand muesli—how much sample do I need to take in order for it to be representative and how many of these samples should I take to show that it is reproducible?

It is now obvious to see that, if we take a smaller **sub-sample** from a larger bulk source, then we can easily introduce sampling errors. We have to be careful in defining how the sample is taken, in order for it to be considered representative. This will be the focus of our next chapter. After all, if you cannot take a sample from the bulk or source material that reflects the required properties for measurement then is it worth bothering to do the analysis at all?

1.2.3 Step 3: sample preparation

What needs to be performed to get the sample into a suitable form for measurement.

In this step the analyst must evaluate the procedures necessary to prepare the sample so that it is ready to be measured. Assuming a suitable representative sample has been acquired, the analyst must ensure that the analyte of interest (or its property) is in a stable form suitable for the proposed measurement technique to be employed. Samples can be solids, liquids, or gases—or mixtures of all three present in a single sample. Whatever their form and composition, the representative sample must be processed so that the property to be measured in the source material is not lost or changed, that is, the sample integrity is maintained. This has to be considered because many measurement techniques require the physical state of the processed sample to be in a particular form, often the liquid state. Furthermore, the method of **sample preparation** is dependent on the later measurement technique to be used, such that it must allow the property to be measured at the level present in the prepared sample.

We shall now go back to our candy samples and consider the analytical challenge we set ourselves: that of identifying some of the natural food additives that are used to produce their colours and determine their concentration. These particular type of candies contain natural colourings from foodstuffs,[5] including lemon (yellow), radish (red), spirulina (blue), safflower (yellow), black carrot (purple), *Hibiscus sabdariffa* (red), and red cabbage (red). These each contain various compounds that give rise to the colours we see. Hence, depending upon the source material, we might be looking for carotenoids, anthocyanins, safflomins, or phycocyanins. Some of these compounds are water soluble and some are not so, to ensure that

---

[4] Food Standards Agency. 1996. The food labelling regulations 1996: guidance notes on quantitative ingredient declarations ("QUID"). Available at: http://www.food.gov.uk/sites/default/files/multimedia/pdfs/quid.pdf

[5] Tolliday, S. 2012. Nestlé confectionary: journey with colours. *New Food Magazine, 13*(6), pp.27–31.

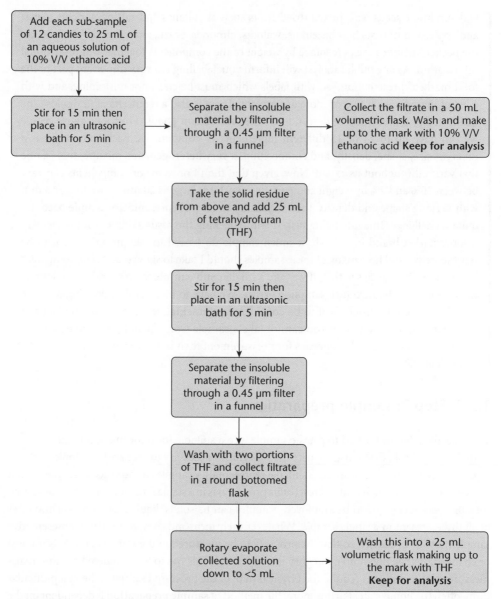

Figure 1.2 Flow chart showing the sample preparation process for the coloured candies sampled in Example 1.1

we extract all the natural colourants from each candy, we will take separate sub-samples, each containing 12 candies of one colour and take them through a two-stage extraction process, as shown in Figure 1.2.

1.2.4 Step 4: measurement

Identifying the most appropriate measurement technique to use, the conditions to apply, and generating the basic data.

In this step the analyst identifies details of the techniques and methods to be used in the measurement process and then uses them. This includes the particular instrumentation, along with any operating conditions and data acquisition procedures to be employed. The analyte to be measured is obviously the target here so a **qualitative** measurement is necessary to correctly identify it and a **quantitative** measurement to determine how much of it is present. During this process the analyst must consider the respective questions of **selectivity** and **sensitivity**. Sensitivity is the change in signal per unit change in amount of analyte, which influences how low a concentration can be detected (the other factor is noise, as described in Chapter 9). The more selective the technique is in targeting the analyte, the better because this often reduces interferences in the measurement. Some techniques are more sensitive than others so it is important to identify whether the technique chosen is sufficiently sensitive to measure the analyte in the prepared sample and that it has a suitable linear range. Validation, using a suitable reference or **certified reference material** (CRM), would normally be included in this step.

How this step is performed is covered in detail in Chapters 4 and 5.

1.2.5 Step 5: evaluation

Taking the data from the measurement step and turning it into values.

In this step the data acquired in step 4 is turned into values which are useful, that is, can be used for appraisal, to help solve the original problem. The corrected values would be worked out, for any dilution steps involved. Hence, the measurement data would be worked back to that property expected in the original representative sample, be it a solid, liquid, or gas. An evaluation of the magnitude of possible interferences and uncertainty in the methodology should also be considered at this stage, because as we have noted, no analytical technique is interference-free! This is discussed in more detail in Chapters 5, 6, and 8. At the end, you should have the relevant information about the property or analyte in the original source or bulk material.

1.2.6 Step 6: analytical information

In this step the analyst looks at the whole process that has been undertaken, the original problem to solve, the requirements of the analysis (why and what!), and critically evaluates it by asking 'Have I completed, with suitable statistical confidence, what I set out to measure and to evaluate?' If it is necessary to compare the set of values calculated for the samples with any regulatory or guidance values then the analyst would do so using appropriate **statistical tests** and taking into account the limitations imposed from the evaluation section. The identification of any patterns that arise from the sample measurements (spatial and/or temporal changes) would be considered here. Where there are limitations, for whatever reason (e.g. sampling errors, accuracy, precision, interferences, etc.) then the **fitness for purpose** should be critically discussed and put in context with regards to the whole analytical process. This is discussed in Chapters 7–9.

All of the steps discussed are shown in schematic form in Figure 1.3 to demonstrate the direction or flow of the analytical process.

If we now return to the candy question and consider the original question: 'how much natural food colourant is present in each of the eight different candies?', then it is instructive to note that we would need to have taken well over ½ kg of sample from over 18 packets in order to acquire some 40 samples or increments for analysis. Five replicates of the eight colours are represented and there are 12 candies in each replicate sample. These increments were each extracted by two methods known to dissolve natural colourants and the types and quantities of colourants present are measured, back-calculated in the original samples, and compared with standards and regulatory requirements using statistical tests.

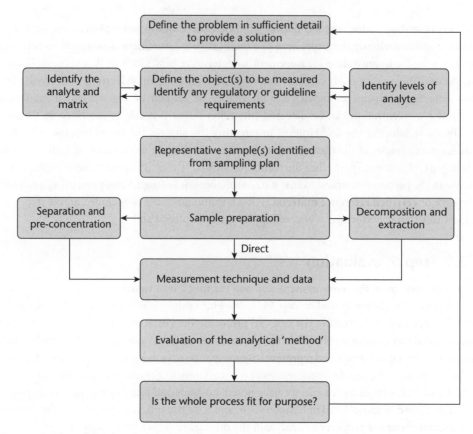

Figure 1.3 Flow chart showing the overall analytical process

1.3 Questions, questions, questions!

Before we look at the various stages of sampling through to evaluation of data in more detail, in the following chapters, it is worth returning to our original step in the analytical approach once more; that of defining the problem and the object being examined.

We have looked at the 'what' and 'why' questions before and asked ourselves 'what are we measuring and why are we measuring this?' Clearly, the background information and the setting of the scene is important and should be known in sufficient detail to allow the third question to be answered: do the two answers just given actually make sense, that is, are they logically linked? This may seem a strange question to ask but actually, it is a critical one.

Before going further, we should settle on an agreed term for what is to be measured. Consider the case where a water sample from a fish tank is sent to an analytical laboratory with a request to measure the nitrate level. In this case the 'what' is nitrate and we refer to this as the analyte. This is a term which will recur throughout this book so it is well to lodge it firmly in the memory now.

The analyst then asks the 'why' question, why do they want nitrate levels measured? The answer: because a number of exotic tropical fish have died while in this water. The analyst asks the why question again, why do they suspect nitrate is a problem (toxicant)? The answer: because a quick check by the fish tank owners, on the total nitrogen content of the water, found it to be about 5 mg L^{-1}. The analyst now performs a quick mental calculation and notes that the total NO_3^- content possible from 5 mg L^{-1} total nitrogen will only be just over 22 mg L^{-1}, insufficient to cause a fatal environment.

However, the analyst knows that if the total nitrogen content were present as ammonium (NH_3/NH_4^+) or nitrite (NO_2^-) then this could create a much more toxic environment and would be a more logical measurement to make. In addition, the presence and levels of NH_3/NH_4^+ and NO_2^- would also inform the exotic-fish keeper about the conditions that create the toxic environment and what needs to be changed to stop this happening again. The effect would be a **value added** measurement, and all because the right questions were asked. It is therefore important to identify whether you will need a **quantitative** and/or **qualitative** measurement in solving the problem (these concepts are discussed again in Chapter 5).

ACTIVITY 1.1 The right questions?

You are a public analyst and a member of the public has just asked to see you. When they walk into your office, they are carrying a capped plastic bottle containing a clear, colourless liquid. Write down what you think are the 10 most important questions you should ask, after they say to you: 'Is this water OK to drink?'

When you have completed this activity, study the feedback given at the end of this chapter.

ACTIVITY 1.2 Levels of nitrate, nitrite, and ammonium

Using the above scenario as a case to study, calculate the actual maximum possible individual levels of nitrate, nitrite, ammonia, and ammonium present in your fish tank water given that the total nitrogen content is (a) 5.0 mg L^{-1} and (b) 7.0 mg L^{-1}. How do these values compare with the guideline values reported in drinking water and in river water? As a starting point for drinking and river water values, you might look to the links: http://www.dwi.gov.uk/consumers/advice-leaflets/standards.pdf and https://www.water-research.net/index.php/nitrate

When you have completed this activity, study the feedback given at the end of this chapter.

We can see that the scenario relating to each case should be presented in as much detail as required to allow a logical solution to be formulated. This means asking questions until the problem is defined. Where there is an important legislative, regulatory, or guidance value involved or associated with the scenario's evaluation, then these values should be stated at this first step or stage. After all, these values will form part of the reason for the analysis and, as we shall see, they will be used in other steps as well.

The fourth question is 'how'. How should I go about tackling this analytical problem? There is an old expression, 'there is no need to re-invent the wheel'. In many cases, you will find that a particular analytical method has already been developed by someone else and published in the scientific literature. Numerous methods are available but **validated methods**[6] are

[6] A standardized or officially recognized approach employing reference materials and/or techniques.

preferred (see Chapters 6 and 9). Even if there are many steps involved in solving an analytical problem, you may find that in some cases an exact match is published which provides your full requirements (particularly those which are considered validated). This is often because the regulatory frameworks previously mentioned are already in place.

It may also be that you have to combine more than one method in order to build up your analytical process to allow a solution to your problem (see Chapters 6 and 9). Hence, some elements of design and modification can come into your answering the fourth question. Of course, when something really new comes along then you may have to consider each of the steps from scratch! But that is what makes the analyst's knowledge, understanding, and experience so important.

 PROBLEM 1.1

Definition of terms

You will have come across a great deal of terminology while reading through this chapter, as yet not fully defined. Every science has its own jargon which makes it a little like another language. This is particularly noticeable when two or more experienced scientists discuss their particular topic area. With this in mind, you are required to define the following words and phrases in order to see just how important they are to the analytical chemist. [Hint: An internet search can help here but note that sources from both the worldwide IUPAC book and the European (Eurachem) guide on analytical chemistry may be of more focussed help.]

When you have done this, compare your findings with the feedback given at the end of this chapter.

- (a) Analyte
- (b) Representative sample
- (c) Regulatory level
- (d) Guidance level
- (e) Test sample
- (f) CRM
- (g) Validation
- (h) Matrix
- (i) Calibration
- (j) Confidence limit
- (k) Accuracy
- (l) Precision
- (m) Interference
- (n) Errors
- (o) Tolerable intake
- (p) Detection limit
- (q) Limit of quantitation
- (r) Sensitivity

(s) Standards

(t) Methodology

(u) Separation

(v) Qualitative

(w) Quantitative

 PROBLEM 1.2

Analytical scenarios

Below, over 40 scenarios of analytical problems are presented for you to choose from. Try to pick one that you feel will be interesting. For this extended activity you will be required to solve the problem in brief form for each of the six steps. Each expands upon the title of the scenario and demonstrates a part of the solution. You can formulate your own storyline based upon the title, or look up published sources, to provide the case to be solved. The end result should be a logical sequence of steps, each step having its own heading, that reads like a scientific story and demonstrates the solution to the problem.

1. A sewage problem: urea in a coastal seawater sample (EA)

2. A food problem: mercury in tuna (FSA)

3. A nutrition problem: Se in our diet (FSA)

4. A metals impurity problem: sulfur in pig iron/blast furnace (industrial)

5. An air pollution problem: H_2S emission from a flue stack at a coal power station (energy generation)

6. A clinical problem: high sodium in a blood sample (hospital trial with new drug)

7. A pharmaceutical problem: steroids (prednisolone) level in a tablet, contaminant

8. A metals production problem: manganese levels in a stainless steel sample

9. A mining problem: arsenic levels in an ore sample (industrial and EA)

10. A soil problem: a pesticide (Gamazene/HCCH) residing in soil at a farm

11. A sediment problem: tin levels in a marina-dredged sample (marine and EA)

12. A biological plant problem: a copper-based fungicide in tea leaves (agricultural industry and FSA)

13. A petroleum problem: Pb in a sample of petrol from abroad (Customs and Excise—now HMRC)

14. A health products problem: silicon in a toothpaste sample (industrial)

15. An occupational accident problem: boron-containing dust in air at a factory after a fire in a $LiBO_4$ flux tank (occupational health, HSE)

16. A nuclear monitoring problem: uranium oxide in a fuel rod

17. An environmental health problem: platinum group metals from road dust in a city centre

18. A plastics problem: plasticizer levels in a plastics sample (PVC manufacturer)

19. An occupational hazard problem: Pb fume levels above a float-soldering bath (occupational health, HSE)

20. A waste problem: levels of NH_3/NH_4^+ in a landfill leachate (WWT plant problem)

21. Polychlorinated biphenyls (PCBs): the presence of PCBs in contaminated ground from beneath a disused electrical sub-station (EA)

22. The level of $CaCO_3$ (a filler) in a loaded plastic: the quantity of filler affects the properties and is a quality control problem (industrial)

23. Arsenic in a paint sample: a 200-year-old stately home with green coloured main hall walls believed to have been painted with copper arsenate

24. The spillage of milk from a dairy farm to a local river: the practice of hosing-out of the cow's milking parlour each day and collection of all wash-waters in an adjacent lagoon is suspected (DEFRA and EA)

25. To determine a measure of the bioavailability of lithium, from the compound lithium carbonate, as used to treat sufferers from bipolar disorder (MRC)

26. Food additive E110 (Sunset Yellow) is suspected as being the colouring agent, present in high concentration in a child's confectionary but not disclosed (FSA and TS)

27. Levels of silver (Ag) in a fabric: silver compounds are used as a biocide to treat fabrics which may have to be worn in harsh environmental conditions (TS and MoD)

28. The levels of chloride and chlorine (free) present in a public swimming pool (Environmental Health and PA)

29. An occupational hazard problem: Mercury fume/air-borne levels within a dentist's surgery (Occupational/Environmental Health/HSE)

30. The levels of certain PAHs (anthracene etc.) in cigarette smoke and the effectiveness of filters (TS and PA)

31. The levels of arsenic and cadmium in allotment soil used to grow edible produce (EA and PA)

32. Bisphenol A in the diet: levels from polycarbonate (PC) baby bottles as the most prominent role of exposure for infants, and/or canned food for adults and teenagers (FSA and PA)

33. The identification and measurement of carbon monoxide levels from a suspected faulty gas-fired heater (Metropolitan Police/Crown Prosecution/HSE)

34. The measurement of levels of antimony in fruit juices and the PET bottles that contain them, as sold to the public (HSE, FSA, and PA)

35. The determination of polybrominated diphenyl ethers (PBDE; flame retardants) in household furnishings and dust, use of which, for some, has now been banned (HSE, TS, and PA)

36. The levels of lead, cadmium, and arsenic in a 'dolomite' sample used in the preparation of Ca/Mg food supplement pills (FSA)

37. The determination of a vitamin (A or C or D or E) levels in a medicinal preparation for the treatment of patients with inflammatory bowel disease (MHA)

38. The measurement of silicon and silicone levels in blood obtained from humans who have undergone implantation with silicone-gel-containing prostheses (HSE)

39. The determination of aluminium and aluminosilicate levels in the preparation of gloss paper (QC/QA)

40. The determination of potentially toxic element contamination in macroalgae (seaweeds) consumed by 'foraging' members of the public for personal or livestock consumption (DEFRA, FSA, and PA)

41. The determination of pesticide and growth-modifier residues on fruit and vegetables, e.g. Alar on apples (FSA, PA, and DEFRA)

42. The determination of iodine/iodide levels in macroalgae (seaweeds) consumed by 'foraging' members of the public for personal or livestock consumption (DEFRA, FSA, and PA)

43. The measurement of tributyltin, an anti-fouling agent added to paints and surface preparations of hulls of ships, from a marina-based seawater or sediment sample (EA, TS, and PA)

Feedback on activities and problems

 FEEDBACK ON ACTIVITY 1.1

The right questions?

You are a public analyst and a member of the public has just asked to see you. When they walk into your office, they are carrying a capped plastic bottle containing a clear, colourless liquid. Write down what you think are the 10 most important questions you should ask, after they say to you: 'Is this water OK to drink?'

Details are important here and the following selection of possible questions will help to define what is required in this case. **Safety** and **risk** must be your first consideration in every analytical problem and you must be aware of what you are being asked to handle. This should be identified from the questioning as early as possible [see http://www.hse.gov. uk/risk/controlling-risks.htm **for guidance on risk assessment process and the document guide:** http://www.hse.gov.uk/pubns/indg163.pdf].

1. Where was the [water] sample actually taken from? This leads into a whole series of other questions which can overlap with those below. The sample may be from the inside or outside of a building or away from any building and in the open (e.g. river, outside tap, allotment source, bore water, land drainage, etc.).

2. Was it from an identified drinking water source?

3. Who supplies this source of water (public or private supply)?

Some water sources are considered as potable water supplies and would be tested by the local authority to check that they are suitable for human consumption. An analyst would sample from that source, knowing it was their only drinking water supply. If the above were not the case other questions might need to be asked:

- You may ask them 'Why do you think it needs to be tested?'
- Describe how the sample was taken
- When was the sample taken?
- Who took the sample?
- What was in the sample bottle beforehand?
- Was anything added to the sample taken?

FEEDBACK ON ACTIVITY 1.2

Levels of nitrate, nitrite, and ammonium

Using the above scenario as a case to study, calculate the actual maximum possible individual levels of nitrate, nitrite, ammonia, and ammonium present in your fish tank water given that the total nitrogen content is (a) 5.0 mg L^{-1} and (b) 7.0 mg L^{-1}.

The activity asks you to calculate the maximum possible individual levels of nitrate (NO_3^-), nitrite (NO_2^-), ammonia (NH_3), and ammonium (NH_4^+) present in your fish tank water given the total nitrogen (N) content. Simple proportion should provide a solution knowing the molar masses of each:

$$\text{Molar mass of:} \quad \begin{aligned} N &= 14 \text{ g mol}^{-1} \\ O &= 16 \text{ g mol}^{-1} \\ NO_3^- &= 62 \text{ g mol}^{-1} \\ NO_2^- &= 46 \text{ g mol}^{-1} \end{aligned}$$

For total N = 5 mg L^{-1}

$$\begin{aligned} \text{Mass concentration of N} \quad &= 5.0 \text{ mg L}^{-1} \\ &= 5.0 \times 10^{-3} \text{ g L}^{-1} \\ \text{Molar concentration of N} \quad &= 5.0 \times 10^{-3}/14 \text{ mol L}^{-1} \\ &= 3.57... \times 10^{-4} \text{ mol L}^{-1} \\ \text{Mass concentration of } NO_3^- \quad &= 3.57... \times 10^{-4} \times 62 \text{ g L}^{-1} \\ &= 2.2... \times 10^{-2} \text{ g L}^{-1} \\ &= 22 \text{ mg L}^{-1} \end{aligned}$$

which is also achieved using the simple ratio method: 5.0 x 62/14; = 22 mg L^{-1}

By performing similar calculations for the other species and tabulating the results we arrive at:

	Concentration/mg L^{-1}	
Total N	5.0	7.0
NO_3^-	22	31
NO_2^-	16	23
NH_3	6.1	8.5
NH_4^+	6.4	9.0

 FEEDBACK ON PROBLEM 1.1

Definition of terms

You will have come across a great deal of terminology while reading through this chapter, as yet not fully defined. Every science has its own jargon which makes it a little like another language. This is particularly noticeable when two or more experienced scientists discuss their particular topic area. With this in mind, you are required to define the following words and phrases in order to see just how important they are to the analytical chemist.

Defining a particular analytical word or phrase can give rise to some differences, depending upon the source employed. It is therefore important to use a validated source for this task.

Two primary sources of such definitions are provided by IUPAC[7] and Eurachem.[8] For example, compare your definitions with the examples chosen below and slowly build up your own glossary of terms which are based upon these important primary sources.

- Analyte: 'The entity (e.g. atom, molecule, ion, functional group, etc.) being investigated'.

- Representative sample: Here we have a phrase based upon the term 'sample', that is, 'a portion of material selected from a larger quantity of material'. The term needs to be qualified and hence it is possible to talk about a 'bulk sample', 'test sample', or 'primary sample' and in this case 'representative sample', which is defined as 'a sample expected to adequately demonstrate the properties of interest of the parent material'. A representative sample may be a **random sample** or, for example, a stratified sample, depending upon the objective of sampling and the characteristics of the population. The degree of representativeness of the sample may be limited by cost or convenience. These points are discussed in our chapter on 'sampling'.

- Validation: The process of determining the performance characteristics of a method to establish whether the analytical results obtained will be fit for their intended purpose. This is discussed in more detail in Chapter 9.

- Calibration: The measurement of standard solutions or materials in order to construct a calibration curve.

- Accuracy: Closeness of the agreement between the result of a measurement and a true (or accepted) value of the measurand (where measurand is the particular quantity subject to measurement).

- Sensitivity: The change in instrument signal (or response) per unit change in amount of analyte; in practical terms this is given by the slope of the calibration curve. If the curve is in fact a 'curve', rather than a straight line, then of course sensitivity will be a function of analyte concentration or amount. If sensitivity is to be a unique performance characteristic, it must depend only on the chemical measurement process, not upon scale factors.

- Qualitative: Analysis in which the identity of the sample and/or presence of the analyte is determined.

? FEEDBACK ON PROBLEM 1.2

Analytical scenarios
Below, over 40 scenarios of analytical problems are presented for you to choose from. Try to pick one that you feel will be interesting. For this extended activity you will be required to solve the problem in brief form for each of the six steps. Each expands upon the title of the scenario and demonstrates a part of the solution. You can formulate your own storyline based upon the title, or look up published sources, to provide the case to be solved. The end result should be a logical sequence of steps, each step having its own heading, that reads like a scientific story and demonstrates the solution to the problem.

[7] International Union of Pure and Applied Chemistry (IUPAC), Analytical Chemistry Division, Commission on Analytical Nomenclature. 2017. *IUPAC Gold Book*. Available at: http://goldbook.iupac.org/indexes/index-alpha.html
[8] Barwick, V.J. and Prichard, E. 2011. *Eurachem Guide: Terminology in Analytical Measurement: Introduction to VIM 3*. Available at: www.eurachem.org/index.php/publications/guides/terminology-in-analytical-measurement

With over 40 different scenarios to choose from, the specific details of the full solutions which you might come up with will vary considerably. It is helpful to consider this task as one that is ongoing and develops throughout the chapters covered. There will be an increase in the accuracy of your solution to the problem as each step is considered in more detail. However, it is possible to use a check list allowing each scenario to be assessed against a set of criteria for the first four steps:

1. Have I suitably defined the problem by asking the right questions, allowing me to identify both the reason for the analytical measurement (why) and what it is I need to measure (what, analyte or property)?

2. Have I identified what I need to do in order to acquire a suitably representative sample from the source (e.g. bulk material) and if necessary keep it stable, in order to allow the later processing of the sample and analytical measurement (analyte or property) to be made?

3. Have I identified what I need to do to the sample in order to allow the analyte or property present in the sample to be in a suitable form for measurement?

4. Have I identified what technique (instrumental or chemical) is suitable to measure the analyte or property now present in the processed sample in terms of sensitivity, detection, selectivity, and specificity?

2

Sampling

Learning outcomes

Once you have read through this chapter, worked through the activities and problems, and reviewed the feedback, you should be able to:

1. distinguish between a homogeneous and a heterogeneous sample;

2. understand the process for obtaining a representative sample from solids, liquids, and gases;

3. create a sampling plan to obtain a representative sample;

4. state details of a variety of methods used for collecting samples;

5. understand the importance of sample preservation.

2.1   Introduction

Imagine for a moment that you are working for an analytical laboratory and your boss has asked you to 'take some samples' from a polluted site. The site to be sampled includes part of a 1 ha crop field (1 ha = 10,000 m^2, about the size of a football pitch), where an unknown, very slightly volatile liquid pollutant has been spilt and is slowly soaking through the soil. The crop field also sits alongside a fast-flowing river. The pattern of contamination is slowly changing as the pollutant seeps through the soil, and there are also health and safety, legal, as well as analytical issues to consider.

Despite the complexity of this nightmare scenario you should be able to make a start by correct identification of the problem, as described in Chapter 1. So, leaving aside the health and safety issues **until question 1 below has been answered** [see http://www.hse.gov.uk/risk/controlling-risks.htm **for guidance on risk assessment process and the document guide:** http://www.hse.gov.uk/pubns/indg163.pdf], you should be asking the following questions:

1. What is this unknown volatile liquid that has been spilt?

2. How far has it spread?

3. How much of it is present?

In order to start answering these questions it will be necessary to take samples from the site for analysis. However, this simple statement in fact hides a plethora of complexity because the critical step, on which all subsequent actions depend, is the process of **sampling**. Before you can even consider how to perform the analysis it is necessary to establish how to take samples safely from the site in a way that is representative of the site itself, a process called **representative sampling**.[1] Provided the samples taken are representative in relationship to the property to be measured from the bulk or source material, this part of the sampling process can be described as being **fit for purpose**.

A **representative sample** is generally considered to be 'a sample resulting from a sampling plan (see Section 2.2) that can be expected to adequately reflect the properties of interest of the parent population'. The 'parent population' refers to the **source** or **bulk material**, in this case the contaminated soil and crop on the 1 ha site.

The analyte is rarely evenly distributed about the bulk material. The sample itself can be **heterogeneous** in nature, and this can be spatial or temporal (i.e. varies with space and time). This gives rise to a problem because we would normally wish to obtain a **homogeneous sample** which is representative of the bulk material. Reconciling this conflict goes to the heart of obtaining a representative sample, as we will see.

. . . .

2.2 **The sampling plan**

The bulk material to be sampled can be a solid, liquid, gas, or mixed phase; it can be homogeneous (e.g. a single liquid) or heterogeneous (e.g. a landfill site); it can be finely divided (a powder) or a single entity (a steel ingot); it can be static (a drum of industrial chemical) or dynamic (a flowing production stream). It can be any combination of these, and the method used to take representative samples will be different in each case. As you can see, the possible combinations are endless with an equally diverse number of possible solutions. So, in order to simplify things we will consider a limited number of scenarios which reflect general principles which we can use to come up with a **sampling plan**.

Let us consider a simplified version of the problem we were given originally in order to identify some of the issues that are likely to arise. Consider the grid in Figure 2.1 as an idealized version of the 1 ha crop field in our original example.

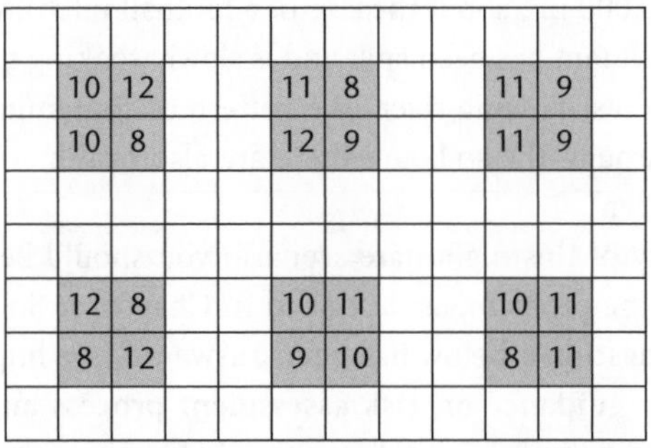

Figure 2.1 Idealized representation of a field showing a theoretical distribution of a component (shown in blue)

---

[1] Representative in relationship to the property to be measured from the bulk or source material.

The crop field has been divided into a grid from which the following important conclusions can be drawn:

- The pollutant is distributed in the blue squares at an overall average concentration of 10 in each set of four adjacent squares, that is, the values in each set of four adjacent squares total 40 (e.g. 10 + 10 + 12 + 8 = 40), so the average is 10.

- The minimum number of blue square samples that must be taken in order to obtain an average of 10 is four.

   a. Taking just one blue square sample will give you a value in the range from 8 to 12.

   b. Taking two blue square samples in any direction (up, down, or diagonally) will give you an average value in the range from 8 to 12 (try it and see).

   c. Taking three blue square samples will give you an average value ranging from 9.33 to 10.67.

- Taking any number of samples from the white squares will give you an average value of 0.

From this simple example we can see that both the **number** and **location** of the samples taken determines whether you obtain a representative value for the component **distributed** in the blue squares.

ACTIVITY 2.1 To help illustrate the problem of taking a representative sample

For the purposes of this experiment you will require a large bag containing around 1000, ~5 mm sized inert coloured beads which will be used to represent the different components in a sample.[2] The beads can be of various colours but one of the colours is chosen to be the focus of the sampling experiment which we will consider to be an active ingredient (or contaminant) depending upon the scenario chosen. It is noted that the beads should all be harmless. The beads should all have the same size and mass and therefore their percent by number, percent by weight, and percent by volume are all identical. It is often simpler to consider just two colours where one colour represents the main sample matrix (e.g. white beads), which for example can be a harmless formulation used to make, say, antihistamine tablets, while the other colour (e.g. red beads) represents, say, the pharmacologically active ingredient, chlorpheniramine maleate, which must be present at a fractional concentration of 0.10 ± 0.01, that is, one 'active' (e.g. red) bead for every 10 beads in the bag (10%; 9 white beads).

You will require 10 mL, 20 mL, and 50 mL plastic beakers and a large tray.

   (a) Mix the beads up in the bag then take a random sample with the 10 mL plastic beaker provided by filling just to the top of the beaker; do not overfill.

   (b) Dispense your sample into the tray and record the count of each colour of bead.

→

---

[2] Based upon the work of Vitt, J.E. and Engstrom, R.C. 1999. Effect of sample size on sampling error: an experiment for introductory analytical chemistry. *Journal of Chemical Education*, 76(1), pp.99–100 and F. Cheng, University of Idaho.

(c) Return all the beads to the original bulk sample bag.

(d) Repeat steps (a) to (c) four more times.

(e) Then, repeat steps (a) to (d) twice more, but once with the 20 mL beaker, and once with the 50 mL beaker.

(f) You will now have a total of 15 sets of counted samples, five for each sampling size.

(g) Be sure to keep careful written experimental notes, of both your observations and the data, preferably in tabular form.

(h) Make a copy of your group's data and compare this with other groups' data as described in the feedback at the end of the chapter.

(i) Be sure all beads are back in the original bulk sample bag before passing to the next group to use.

(j) Process the results, as outlined in the 'feedback for Activity 2.1'.

The preceding scenario involving a pollutant and Activity 2.1 clearly illustrate that we need to control how we sample. In order to do this we need to present a **sampling plan**. The sampling plan is a detailed procedure undertaken to acquire the sample(s) and identifies the critical steps in this methodology as follows:

1. Sampling location

    (a) from where to take samples;

    (b) will samples need to be taken at the surface or at depth or both in the case of solid and liquid sources?

    (c) whether the samples taken are 'random' or selective/judgemental;

    (d) should the samples be composite or separate?

2. Sampling frequency

    (a) will the output from a source be changing periodically, fluctuating between high and low values over time?

3. Sample size

    (a) how much sample to take (and how many)?

    (b) will there be enough for all analyses?

4. Sampling methodology

    (a) how to actually take the samples and in a safe manner (**health and safety, risk assessed**);

    (b) how will the samples be stored and preserved so that the parameter to be measured does not change form or concentration?

    (c) if sampling solid materials, will that include taking core samples (i.e. a cylindrical section of the source material obtained by drilling with special apparatus)?

    (d) if separate phases are present, how do they contribute to the analytical measurements being made (e.g. suspended solids, particles, in a river water sample, particulate matter in an air sample, or the gas associated with a soil or water sample, etc.)?

(e) is any mixing of the source material required prior to sampling on site or filtering (or even centrifugation) for later sub-sampling;

(f) how are gas or atmospheric samples taken so they can be considered representative, from where, for how long, and what total volumes are to be sampled? Are filters to be used and if so what size should they filter down to (what 'pore' size is required to trap the particles)?

The analyst needs to provide full details for the sampling plan so that, if required, others can reproduce what was undertaken and limitations can be identified in the later evaluation step.

. . . .

2.3 Sampling location

How you approach your sampling, whether it be the sampling of solids, liquids or gases will depend upon the problem presented and how it can be solved. However, some general procedures may be identified for various scenarios.

2.3.1 Random or selective sampling?

Continuing with another agricultural example, consider the case of two crop fields being prepared for planting maize. Each is approximately 1 ha in size. One field requires its nitrogen (N) and phosphorus (P) levels to be measured in the soil prior to adding fertilizer and planting. Fertilizers aid growth but they are also costly. This treatment is referred to as **dressing** with fertilizers. In this example, we are dealing with a bulk soil source. In theory one might expect the N and P levels to be within a certain range throughout the field because it is dressed each year with the required amount of fertilizer, planted, and then harvested. The exposed field then weathers so that the levels of N and P decline to a value similar to that identified from previous years' analyses. Assuming that the dressing and density of planting is evenly distributed each year and that the landscape and weathering effect is similar over the hectare site, then one approach might be to consider dividing the total field area into a sequentially numbered grid. A suitable number of randomly selected cells from the grid, say 10% of the total field area, could then be identified using a simple random number indicator programme in order to decide from which cells to take the samples. An example of such a grid is shown schematically in Figure 2.2.

	1	2	3	4	5	6	7	8	9	10
1		X								
2					X					
3			X							
4		X								
5							X			
6										X
7				X						
8								X		
9	X									
10						X				

Figure 2.2 Schematic diagram of one possible random sampling grid for a field

A number of further considerations will influence the final sampling plan:

1. The size of the field will have a bearing on how many samples to take. This will depend upon the level of accuracy and precision required in order to satisfy the objectives of the analysis.

2. The heterogeneity of the field.

   (a) There may be different soil types that are present which will influence the distribution of the analyte(s) and the heterogeneity of the bulk matrix.

   (b) Non-uniform application of fertilizer.

   (c) Different past usage, previous crops, livestock confinement.

   (d) Varying landscape.

3. Composite or incremental analysis? If one simple value is required from the analysis to guide the farmer, then a suitably prepared, mixed set of samples (where our individual samples are referred to as increments) could be collated to provide a single **composite** test sample for analysis. It is also important to understand that while this presents an average value for the entire field, it provides no information on the variation of values or their distribution; information which may be beneficial and only comes with individual **increment** analyses.

As a consequence of the above, it may be necessary to sub-divide the field into separate areas for collection of representative samples. Two main approaches are considered: **systematic** and **random**. The term 'systematic' relates to being methodical, regular and orderly. The term 'random' relates to being non-selective and unbiased. Further, adjacent sites can be **aligned** or **unaligned**. Examples of these combinations are shown in Figure 2.3.

In Figure 2.3a, a totally random approach to sampling has been taken. However, the possibility of large areas being missed is evident. A systematic grid approach (Figure 2.3b) allows the whole area under study to be sampled if required and it is possible to introduce a random element into the sampling within the grids, as shown in Figure 2.3c. A further refinement is the use of the **herringbone** sampling design, as shown in Figure 2.3d, which offers many benefits in our attempts to acquire representative sampling from the whole site[3].

2.3.2 Composite sampling

If there is only the requirement of a simple value or limited set of values for an analyte or property of interest to be acquired from a bulk source material or large area, then it may be possible to collect a suitable set of sub-samples and bring them together to form a **composite sample** which will be used for analysis or measurement. Some general rules for compositing discrete sub-samples from field samples are given below:

1. Samples should be

   (a) equal in size (to avoid bias);

   (b) from immediately adjacent sampling points, particularly if reducing many sub-samples down to a limited set of composite samples;

   (c) evenly spaced to reflect the distribution;

   (d) composited laterally, that is, from the same depth.

[3] See Ferguson, C.C. 1992. The statistical basis for spatial sampling of contaminated land. *Ground Engineering*, June 1992, pp. 34-38.

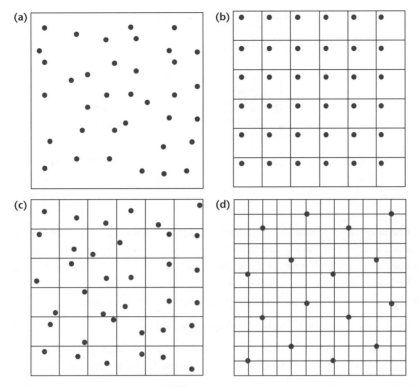

Figure 2.3 Schematic representations of different types of sampling plan: (a) random within entire site; (b) systematic aligned grid with random start; (c) random within grids; (d) herringbone (systematic unaligned with random start)

2. No more than four sub-samples should be included in a first step composite sample. It may be possible to bring together and mix up to four first step composite samples, so as to produce a second step composite sample, and so on.

3. Lines connecting the discrete sub-sampling locations must not intersect with those of other composites.

4. Each composite area must be of similar size and shape, except where not geographically possible.

5. Each discrete sub-sample must be thoroughly homogenized (preferably in the laboratory, rather than in the field) before taking any sub-sample from the composite.

6. Each discrete sub-sample must contribute an equal amount of material to the composite.

Composite samples can be acquired over time and not just spatially: for example, an analyte in the aqueous discharge from a waste water treatment plant where the average daily output to an estuary is required rather than the temporal variability.

2.3.3 Selective sampling

Consider the case where another field is also to be used for growing maize, but has not been used for some time because of a pollution incident that had occurred 5 years earlier. A batch of sewage sludge was used to raise levels of nitrogen, phosphorus, and organic carbon in the soil

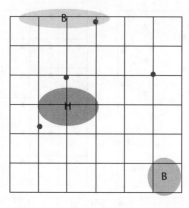

Figure 2.4 A field showing the hot spot area of metal contamination (H) and the background levels (B). In this case, random sampling (blue dots) has missed the hot spot

but unfortunately it also contained high levels of copper, lead, and zinc. It had been decided at that time to leave the field fallow and to let natural weathering and time ameliorate the effects, but it is now necessary to determine the levels of metals within the field prior to planting. Only part of the field had been contaminated.

In this example, two sequential analytical processes can be identified:

1. to determine if the metal levels exceed guideline values;
2. if the metal levels are below the guideline values, then what are the levels of N and P in the field's soil in order to prepare for addition of fertilizer (dressing) and planting?

Previous examination of the field at the time of the incident had identified the location and concentration of the metals, so the sampling plan can include taking samples from identified areas in the field called **hot spots**, as well as control samples at locations away from the hot spots to determine general background levels in the soil.

This type of sampling approach is called **selective** (or **judgemental**) sampling, in order to distinguish it from random sampling. If a random sampling approach was used here, using a grid or cell system, then it is possible to miss a hot spot, as shown in Figure 2.4.

EXAMPLE 2.1

Mobile when molten, trapped when solid

A mixed, and then poured, 1 tonne batch of molten steel was taken from an 'oxygen converter' and cooled to produce a solid, square cross-sectioned ingot. This 1000 kg batch occupied a volume of ~0.127 m³ (0.356 m × 0.356 m × 1 m long). It is important to know how much carbon and other components are in the steel before it can be further treated. The carbon content should be around 0.5% in the grade of steel produced and different sampling approaches were taken to investigate the levels contained. Three 1.00 kg test samples (each of ~127 mL in volume) were then taken from this batch by drilling 2 cm diameter holes through the whole depth of the ingot, in the middle and near both the ends. The carbon content of the three individual samples was nearly identical (Table 2.1, column 2). The range in % carbon was narrow and small, demonstrating very good precision. However, the composition of individual 10 g test samples taken from shallow drillings along the length of the ingot on its top, bottom, and sides showed greater variation in carbon content (Table 2.1, column 3) with a broader range and therefore poorer precision but also a higher mean value than that from the larger 1 kg samples previously measured. Further individual 10 g test samples drilled below the first series were more consistent and showed slightly less variation in their range of carbon concentration and hence greater precision (Table 2.1, column 4). Importantly, the mean value from these five samples was much closer to the mean from the large 1 kg samples value. When individual results from 100×1 g drilled samples were pooled (Table 2.1, column 5), the mean value was nearly identical to those of the individual 1 kg samples but the range in values covered was the greatest of all the analyses and demonstrated the poorest precision.

Table 2.1 Carbon content of samples taken from a steel ingot

Sample number	Carbon content/%			
	1 kg increments drilled right through	10 g increments from surface	10 g increments below surface	1 g increments taken randomly
1	0.455	0.479	0.466	0.441
2	0.448	0.489	0.449	↓
3	0.451	0.497	0.471	↓
4		0.463	0.458	↓
5		0.476	0.453	↓
↓				↓
100				0.562
Mean	0.451	0.481	0.459	0.453
Range	±0.0035	±0.017	±0.011	±0.12

Variations in the distribution of the analyte (carbon) can be seen within the steel source being sampled. These variations will be dependent upon the size of the sub-samples taken. The carbon distribution is locked into each defined volume of the solid steel ingot at the time it solidifies. As the number of samples is increased the mean value becomes more representative of the bulk sample. These variations in the carbon distribution can give rise to **sampling errors** and are reflected in levels of both accuracy and precision.

Returning to our terms **homogeneous** and **heterogeneous** and how they relate to the distribution of an analyte in a sample, we might demonstrate this diagrammatically, as shown in Figure 2.5.

We shall see in Chapter 3, which deals with sample preparation, just how critical the homogenization process is to acquiring representative sub-samples for analysis.

2.3.4 Stratified sampling

If we knew our metal contamination in the field was in a particular region but unsure exactly where, then we may wish to focus our sampling in a stratified way. In **stratified sampling**, the area is divided into non-overlapping sub-areas where samples are taken using a particular approach. The approach would be chosen to provide information to help solve the analytical problem. This could include clustering of the samples in a systematic way in the sub-area which contains the contamination, while taking a more random approach in sub-areas that do not contain the contamination.

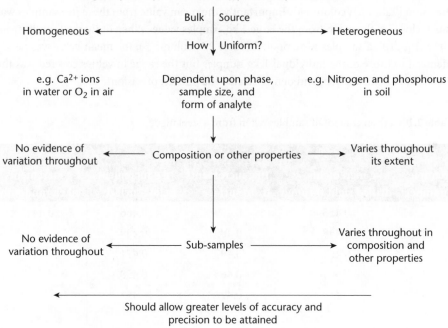

Figure 2.5 Relationship between homogeneity, heterogeneity, distribution of analyte, and type of sample

EXAMPLE 2.2

The detection of hot spots and numbers of samples

The risk of missing a hot spot depends on the size of the plot to be sampled from, the size of the hot spot within this and the number of sampling points. The procedure[4] for determining the number of sampling points required for hot spot detection is based upon detecting **circular hot spots**[5] with 95% confidence (5% risk of missing) using a **square grid** sampling pattern, and the equations for detecting circular hot spots in a contaminated field.

For example, say you needed to calculate the minimum number of sampling points, n, required and the size of grid needed in the sampling pattern, given that there is a contaminant somewhere within a 0.8 ha site. You want to know, with 95% confidence, that you can detect a circular hot spot within this field as small as 10 m diameter. Proceed as follows:

1. Determine or estimate the radius of the smallest hot spot, r (in metres), that needs to be detected. If the diameter of the circular hot spot is 10 m then the radius, r, is 5 m.

2. Calculate the size of the grid, G, using:

$$G = \frac{r}{0.59}$$

(2.1)

---

[4] Gilbert, R.O. 1987. *Statistical Methods for Environmental Pollution Monitoring*. John Wiley & Sons.

[5] Other equations are required for non-circular, elliptical hot spots.

where G is the distance between two sampling points, that is, the grid size of the sampling pattern in metres. The constant, 0.59, in the equation is based upon the hot spot being circular and a confidence level of 95%.

The grid size is therefore: $G = \dfrac{5}{0.59} = 8.5$ m

So, for a square-based grid pattern within the whole sampling area the sampling points are 8.5 m apart.

3. Determine the number of sampling points required, n, for an overall size of sampling area, A (in m²) using:

$$n = \frac{A}{G^2} \tag{2.2}$$

The number of sampling points is: $n = \dfrac{8000}{8.5^2} = 111$

So, the number of sampling points is 111 and the grid size of the sampling pattern should be 8.5 m (sampling points = 8.5 m apart). If the 0.8 ha field was square then for practical purposes you could divide it into an 11×11 grid system (121 squares) and rearrange equation (2.2) to solve for G as follows:

$$G = \sqrt{\frac{A}{n}} \tag{2.3}$$

$$G = \sqrt{\frac{8000}{121}} = 8.1 \text{ m}$$

Now try Problem 2.1 at the end of this chapter.

2.4 Sampling frequency

There are many obvious cases that involve sampling spatially, whether it be in two or three dimensions. However, changes over time result in patterns of variation which have a time-based (temporal) distribution. We were introduced briefly to an example when considering composite sampling, that is, the aqueous discharge from a waste water treatment plant, where the average daily output in concentration of an important analyte is required and not its hourly variability. Taking a representative sample on a time basis depends on the pattern of change of the analyte or property. For example, if a concentration is increasing over time: will it continue to increase, or go down after a certain time; will it just be a transient spike; or will it cycle and show a periodicity, like a sine wave? Changes could be over minutes, hours, days, and, as we can often see in the natural environment, seasons.

The problem is illustrated by the simplified representation of temporal variation shown in Figure 2.6. The analyte value (y-axis) is shown by the continuous blue line against time (x-axis). If you sampled at intervals given by the dashed arrows, then measurement of each incremental sample would give a good estimation of the analyte profile, giving mean, median, maximum, minimum, and frequency information. However, if you took samples every other arrow then the results would be totally different.

The information you gain over time may often be considered as part of a monitoring programme. This can be applied to quality control in industry (e.g. sampling the pharmaceutical pills from a production line) or protecting the aqueous environment (e.g. monitoring industrial discharges).

Some time-based effects can be quite complex in their profile, as quite often occurs where the analyte is associated with liquid or gaseous substances. One rather unfortunate forensic scenario is presented in the example.

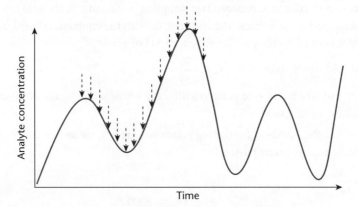

Figure 2.6 Representation of the analyte profile in a temporal sampling scenario

EXAMPLE 2.3

Temporal sampling from a heater flue

A gas-fired (by propane), wall-mounted water heater, used in a second-hand caravan, was suspected of producing high levels of carbon monoxide (CO) after the owners were admitted to hospital with symptoms of carboxy-haemoglobin poisoning. Blood tests confirmed the presence of high levels of CO. The only possible source of CO/CO_2 inside the caravan was the heater because all other appliances were electric. Neither of the owners were smokers and all motorcars were kept in a separate parking area to that of the caravan at the time of the incident.

In this example, the presence of a potentially lethal toxic gas such as CO would require a special gas sampling set-up to allow a safe environment for the test to be conducted for all involved. The presence of the heater in a contained environment, such as a caravan, which restricts diffusion of the gases away from the site has a bearing on any rising concentrations of gaseous products. Hence, the concentration of any CO would be time-dependent; known as a 'temporal' effect. The water heater would be started, without changing any controls or conditions of its operation from those at the time the incident occurred. Samples can be monitored in real time using CO-selective sensors at various locations within the caravan, providing a temporal trace of concentration against time. Alternatively, a suitable gas sampling apparatus may be employed to take known volumes of the atmosphere at selected times and locations within the caravan and to measure the CO levels in the individual samples (increments). The positions of the samplers or the sensors could be based upon a judgemental or stratified approach in order to obtain a representative sample.

2.5 Methods of sampling

The analyst is always required to obtain a suitably representative sample from the bulk source in order to answer a particular question about the bulk material. If the sample is taken using equipment that is inadequate then an error will be introduced. The physical action of taking a sample from the bulk will depend upon the form and presentation of that material. The different phases of solids, liquids, and gases provide different challenges when attempting to obtain a representative sample, and careful consideration must also be given to addressing **sample integrity** when this process is undertaken.

2.5.1 Sampling solids

Taking a representative sample from a solid material is often challenging due to its form and presentation. A mild steel ingot presents a different challenge compared with a river sediment; sampling from a batch of processed pharmaceutical tablets will require a different approach to sampling from a dispensing silo or an ore pile.

A static material sample (e.g. an ore pile) will require a different approach to a moving source (e.g. a conveyer belt of crushed ore). Where a static material demonstrates a three-dimensional variation in its properties, then a representative sample may be taken by a selective sampling approach. The movement of particulate source material which is heterogeneous in size and density will change in presentation over time, that is, particle size, shape, and density will change as the conveyer belt moves along. Additionally, if you shake a sack of multi-sized particulate material then the large particles rise to the top of the sack and the small move to the bottom. Flowing materials, such as sand or grain products, may be sampled from dispensing containers such as silos but possible segregation effects may still require taking composite samples over the dispensing period to provide a more representative sample for measurement.

Any equipment used must itself be **fit for purpose**, so that sampling errors are eliminated and sample integrity is maintained.

Metal samples. If direct methods of measurement are not possible then sampling may come down to use of:

(a) particular drilling equipment made of a material harder than the 'measure of hardness' of the bulk material to be sampled while also not being made of the analyte itself, for example, tungsten carbide drilling of hard metals (provided tungsten is not the analyte);

(b) a **percussion mortar** to sample a metal which is crystalline and fractures easily, once again provided that mortar is made of a material harder than the bulk metal material to be sampled and not made of the analyte.

Powder samples. For relatively narrow range finer particulate materials, as dispensed from silos or stored in sacks and bags, for example, food products, fine chemicals, etc., then a **sampling thief** may be used (Figure 2.7) if a free-flowing or poured sample is not available.

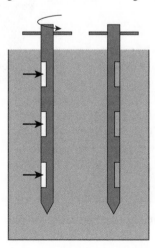

Figure 2.7 Sampling thief used for sampling particulates

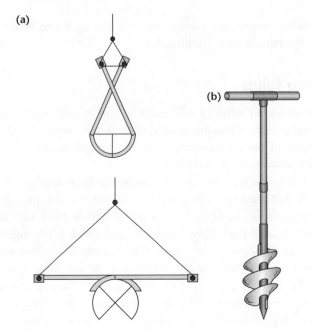

Figure 2.8 Methods for sampling bulk particulates: (a) grab sampler for sampling sediment; (b) auger for sampling soil

Bulk particulate samples. For bulk coarse-to-fine based particulate materials, such as sediments, soils, and ores, then grab samplers, soil and sediment corers, and augers may be used (Figure 2.8).

Often, the main problem with sampling large bulk particulate samples is obtaining a small sample which is representative of the bulk. Methods for sample size reduction include **cone and quartering** and a **spinning riffler** (Figure 2.9). In the former, a cone is made of the bulk material which is then flattened and quartered; the quarters can be formed into cones and further subdivided. However, segregation effects associated with the movement of multi-sized and multi-density particulate material can result, so the spinning riffler is often preferred. In

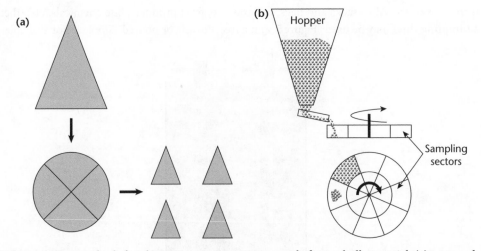

Figure 2.9 Two methods for obtaining a representative sample from a bulk material: (a) cone and quartering; (b) a spinning riffler

this method, a circular holder rotates at a constant speed, and the sample is loaded at a constant rate into the containers via a vibratory chute, which is fed by a mass-flow hopper. This provides more representative sub-samples from a larger, bulk source.

Rock-based materials. Here, one would resort to rock and ore crushers in order to reduce the particle size allowing one of the above samplers to be used subsequently.

Softer-based materials: wood and plastics/polymers. Here, drilling, cutting, and splitting techniques would be used to acquire smaller volumes for analysis.

2.5.2 Sampling liquids

The mobile nature of the liquid bulk material often allows a more representative sub-sample to be taken. However, any changes in physical and chemical character can also occur more rapidly than in a solid. Temporal effects are therefore important so, where possible, *in situ* or **in-line** measurements (i.e. measurements at the source) should be considered. The lower density of such sample types and the low concentrations often encountered means that the volume of sample to be taken must be carefully considered. Flow samplers and sample collectors are available for a variety of liquids which are designed to maintain stability and integrity of the sample and the analyte. These are made from a variety of materials depending on the analyte and nature of the sample, for example, glass, plastic, coated, opaque, sealed, airtight, flushed, etc.

Specialist samplers are available for industry and environmental sampling, dependent upon any standard operating requirement and the type of liquid under investigation. Tube samplers allow both composite and selective depth samples to be taken from tanks and water bodies (lakes, rivers, and ocean). For example, a van Dorn bottle (Figure 2.10a) is used for sampling waters at a particular depth. A glass thief can be used to sample a chemical from a drum (Figure 2.10b) and a splitter sampler is used to sample flow or process streams (Figure 2.10c).

2.5.3 Sampling gases

As with liquids, the mobility and density of a gaseous source can present both temporal and quantity sampling problems, but to a greater extent. Containment and integrity of the sample are also challenges so, as with liquid systems, where possible *in situ* or in-line measurements at source should be considered (e.g. direct air-sampler monitors or gas sensors). Alternatively, there are various approaches that can be taken in the sampling of volatile organics and gases including grab sampling, solid-phase trapping, liquid trapping, headspace sampling, purge and trap, and thermal extraction. Collection vessels include evacuated flasks, gas or liquid displacement containers, flexible plastic containers or gas bag samplers, and even hypodermic syringes. The inside surface coating of all these are carefully controlled to maintain integrity and avoid losses. Active samplers include cold traps and liquid sorbents like Dreschel bottle systems, while solid sorbents like activated charcoal (Draeger) or silica gel can be used.

It is important that the volume of the container and/or the flow rate of the sampling system is exactly known. The flow rate should also be kept constant during the sampling procedure.

Solid-phase trapping: a gaseous sample is passed through a tube packed with a suitable adsorbent (e.g. silica gel, activated carbon, etc.) and the trapped analytes are eluted with a strong solvent. In this approach the gas flow rate is critical for the trapping efficiency. Important parameters are aerosol formation, adsorbent overloading, and irreversible adsorption of reactive analytes. Selective chemical modification of the adsorbent can be useful in improving trapping efficiency, as well as the purge and trap technique.

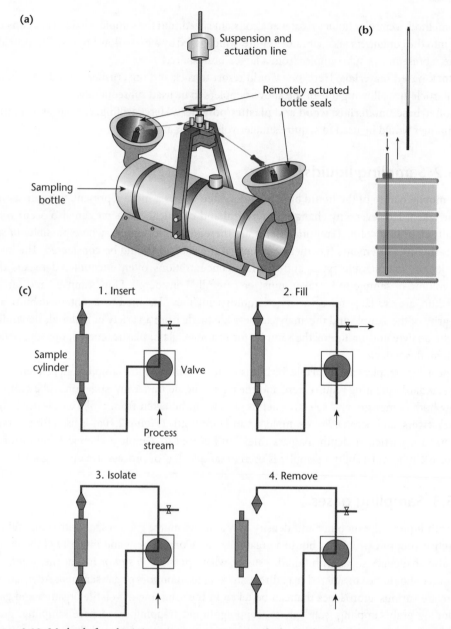

(a)

Suspension and
actuation line

Remotely actuated
bottle seals

Sampling
bottle

(b)

(c)

1. Insert

Sample
cylinder

Valve

Process
stream

2. Fill

3. Isolate

4. Remove

Figure 2.10 Methods for obtaining a representative sample from a liquid: (a) van Dorn sampler;
(b) glass thief; (c) splitter sampler

Volatile organic sampling tubes (VOST) can be used for the sampling of volatile compounds
with a boiling point of 30–100°C. A sorbent (e.g. Tenax TA 35/60) is present and by using a flow
rate of 250–1000 mL min⁻¹, about 5–20 L of gas can be passed through the tube. In most cases
thermal desorption is used to elute the analyte from the sorbent to the measuring equipment.

Liquid trapping: a gaseous sample is passed through a solution, which is a good solvent for the
main analyte gas or analytes contained within it. If the gas itself is not the focus of the sampling
then that may pass through the solvent trap solution unabsorbed. It is important that the flow
rate should be low enough so that no foams or aerosols are created. Complexing agents may be
added to the solvent to aid trapping and the temperature can be lowered for very volatile species.

2.6 Sample preservation

If it is not possible to make a direct measurement on the sample at the point of taking it, then **preservation** must be considered. Once a sample is removed from its original environment then maintaining its **integrity** is critical. No addition to or loss from that sample should occur, and its composition should remain unchanged. Hence, any physical, chemical, and biological processes which might occur are minimized so that the analyte is preserved. The importance of the container used for sample collection and the time between collection and preparation for measurement must be considered.

Some sample preservation techniques and their limitations are shown in Table 2.2. It should be noted that there is a distinction between sample preservation and sample preparation: there are various sample preparation processes that may go on to supersede the preservation step, but these will be considered in the next chapter.

Table 2.2 Sample preservation techniques

Analyte/property in matrix	Containment	Preservation step	Timescale
Metals (not Hg) in soils and sediments	In suitable plastic**, sealed container	Keep cool or freeze and in the dark	As soon as possible if not frozen
Metals (not Hg) in biological solid materials (flora and fauna)	In suitable plastic, sealed container	Freeze and keep in the dark	As soon as possible if not frozen
Hg in soils, sediments and biological solid materials (flora and fauna)	In glass, sealed container	Freeze and keep in the dark	As soon as possible if not frozen
Metals (not Hg) in aqueous liquid samples	In suitable plastic, sealed container	Reduce pH to <2 with HNO_3	<6 months
Hg in aqueous liquid samples	In glass***, sealed container	Reduce pH to <2 with HNO_3	<1 month
Metals (not Hg) in non-aqueous* liquid samples	In glass***, sealed container	Freeze carefully and keep in the dark	As soon as possible if not frozen
Hg in non-aqueous liquid samples	In glass, sealed container	Freeze carefully and keep in the dark	As soon as possible if not frozen
Anions such as SO_4^{2-}, PO_4^{3-}, Cl^-, in aqueous liquid samples	In suitable plastic, sealed container	Freeze and keep in the dark Or reduce pH to <2 with HNO_3	As soon as possible if not acidified or frozen. If acidified, <1 month
Anions such as SO_3^{2-}, PO_3^{3-}, NO_2^-, in aqueous liquid samples	In suitable plastic, sealed container— exclude air	Freeze and keep in the dark Reduce aerial oxidation	If not frozen, as soon as possible

Table 2.2 (*Continued*)

Analyte/property in matrix	Containment	Preservation step	Timescale
Anions such as NO_3^-, in aqueous liquid samples	In suitable plastic, sealed container	Freeze and keep in the dark Or reduce pH to <2 with H_2SO_4	As soon as possible if not acidified or frozen. If acidified, <1 month
Anions such as sulfide, in aqueous liquid samples	In suitable plastic, sealed container— exclude air	Freeze and keep in the dark Reduce aerial oxidation Keep pH >7 if not frozen	As soon as possible if not frozen
Non-volatile organic analyte in soils and sediments	In glass, sealed container but some plastics may be considered	Keep cool or freeze and in the dark	As soon as possible if not frozen
Non-volatile organic analyte in biological solid materials (flora and fauna)	In glass, sealed container but some plastics may be considered	Keep cool or freeze and in the dark	As soon as possible if not frozen
Semi-volatile[+] organic analyte in soils, sediments and biological solid materials (flora and fauna)	In glass, sealed container	Freeze and keep in the dark	As soon as possible if not frozen

* Non-aqueous liquid samples = Organic solvents and similar
** Plastic = PP, PE, PTFE. Take note of contaminants from certain plastics
*** Glass = Note contaminants from glass and care when freezing liquids
[+] For example, PAHs and PCBs

ACTIVITY 2.2 Sampling plan

Explain the procedure you would use to obtain a representative sample of contaminated water from a waste water treatment plant, at the point of discharge into a river. Where else should samples be taken for comparison? [Hint: See the stepwise guide in section 2.2 'The Sampling Plan', and sections 2.5.2 and 2.6, to help give you the brief but important first points to consider.]

Sampling potentially introduces the greatest uncertainty to the analytical process, so it is necessary to acquire a representative sample, whose characteristics reflect that of the analyte in the bulk material or source. The sample taken should also allow you to make a suitable measurement that will help to solve the problem you have identified.

The equipment used to extract and contain the sample must allow the integrity of that sample to be maintained for the period up to the step where sample preparation is undertaken.

In Chapter 8, where we will return to this problem by taking a quantitative approach to determining the **sampling error**, you will learn how to minimize sampling errors, decide how much sample to take, and how many.

? PROBLEM 2.1

Sampling 'battleships'

The game of 'battleships' is an old one. Modern day updates can be found online and even marker boards can be purchased or made.[6] The general game and rules are well known and covered on the internet.[7] Essentially, an area of battle is presented in the form of grid lines, usually square grids and often for simplicity shown as a 10×10 grid where up to five battle vessels of different sizes (can cover more than one square grid) are randomly positioned. You have to locate these by estimating and presenting their position to your opponent, who informs you if you have made a 'hit' or a 'miss'.

In this variant, you are an analyst trying to locate where contamination is present in an area. All you know is that there are five places in a field that are contaminated but you don't know where. Your task is to locate all five and the areas they each cover. However, in this version you will use science and apply the equations used to sample hot spots. This will allow you to construct the grids of the right size and arrangement to enable you to make calculated guesses with confidence.

All you know at the start is the size of the field (area = A), that there are five contaminated places of known size/shape, and that you have to use the equations to help you identify where each of the five hot spots actually are. Use the equations to first construct the right grid size and arrangement before any guesses are made.

For example, a field to be sampled is 1 ha (10,000 m^2 or 100 m × 100 m if square) in size and has five contaminated hot spots of the following sizes:

- 2 of 5 m diameter
- 1 of 10 m diameter
- 1 of 20 m diameter
- 1 of 30 m diameter

You need to locate these five hot spots and the area they each cover to win the game. However, to do this you have to first divide up the whole area into grids that you can sample from. You can calculate how many grids and sampling points you need using the equations used in Example 2.2 and using the smallest dimension possible from the five shapes given, in this case 5 m.

First calculate the grid size: $G = \dfrac{2.5}{0.59} = 4.24\ldots$ m

Then calculate the number of sampling points: $n = \dfrac{10,000}{4.24\ldots^2} = 557$

[6] http://www.hasbro.com/common/instruct/Battleship.PDF
[7] https://en.wikipedia.org/wiki/Battleship_(game)

For a square field this is a 23.6 × 23.6 grid. If we take the closest whole number of 24 then we are talking about a square field with a 24 × 24 grid (576 sampling squares in all), which looks like the area below shaded in grey:

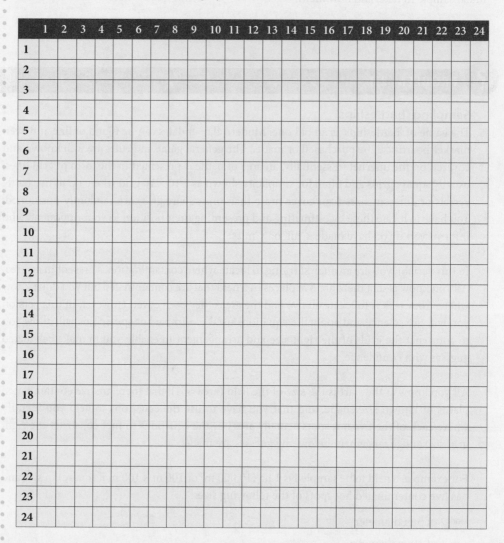

Recalculating the grid size, based upon the whole number of $n = 576$, gives

$$G = \sqrt{\frac{10,000}{576}} = 4.2 \text{ m}$$

Therefore, sampling points are 4.2 m apart. Because the sampling grid size is less than 5 m we should have a greater than 95% chance of actually hitting our 5 m circular hot spot when we land on a square with the hot spot in it.

You can now have 10 analytical sampling guesses to see if you can locate any of the hot spots by choosing 10 sets of x and y coordinates. Your opponent will then try their 10 analytical sampling guesses to see if they can locate any of the hot spots for their own grid set. If one or more of the coordinates you guessed coincides with part of a hot spot then you will be informed by your opponent. The game continues, in turn, your guesstimates and then your opponent's guesstimates, in a sequential manner. As you begin to pick up hits, these will help to inform your next set of guesstimates. The winner will be the first to correctly identify all the contaminant hot spots with all the grids they cover.

Feedback on activities and problems

? FEEDBACK ON ACTIVITY 2.1

To help illustrate the problem of taking a representative sample

1. For each individual sampling, calculate the fraction (f_a) of 'active' beads (i.e. those representing the active ingredient) in the total.

   $$f_a = n_a/(n_m + n_a)$$

   where n_a = number of 'active' beads and n_m = number of matrix beads. The result will be in the range from 0 to 1.

2. For each sample size (beaker size), combine the five individual measurements of fractional 'active' bead concentration to calculate the mean, sample standard deviation, relative standard deviation (RSD), and percentage relative standard deviation for the fractional 'active' coloured bead concentration. The theory and equations required for this are readily available but are also given in Chapter 7.

3. Comment on the values obtained, particularly the way in which the RSD changes with number of beads sampled/volume of sampling beaker.

? FEEDBACK ON ACTIVITY 2.2

Sampling plan
Explain the procedure you would use to obtain a representative sample of contaminated water from a waste water treatment plant, at the point of discharge into a river. Where else should samples be taken for comparison?

Your sampling plan should consider health and safety (risk assessed), the locations (see below), the sampling depth, the volume and the number of samples to take, in order for these to be fit for purpose, that is, to be able to help you solve the problem. Various tests would need to be performed and a sufficient sample should be an overestimate. Consider the timing of the samples (over timescale) and the stabilization of the analyte in order to maintain integrity of the sample, that is, acidified for metals and metalloids, while employing a plastic container, glass for organic compounds and Hg species, keep cold/close to freezing to reduce volatiles, and in the dark to reduce photodegradation effects. Take your selected, suitable pre-cleaned container in which to acquire samples (glass or plastic) and rinse the bottle and cap with the sample collection water at least twice before the final collection. The sample should fill the container to the top and leave no air gap. You should consider taking water samples at the point of discharge (from the exit pipe itself if possible, just where it enters the river). You should sample just after the point of entry to the river and also at selected points further downstream to demonstrate the dilution effect, if necessary. However, it is important that you also sample from the river just before the discharge pipe's point of entry, to act as a control or baseline. This is in order to evaluate the effect of the discharge on the river at that point. The samples are considered selective and judgemental around the contamination site (but it may be required to take some samples, both before and after in a more random manner).

 FEEDBACK ON PROBLEM 2.1

Sampling 'battleships'

Taking our original example of five contaminated circular hot spots of differing dimensions,

- 2 of 5 m diameter
- 1 of 10 m diameter
- 1 of 20 m diameter
- 1 of 30 m diameter

and based upon our now calculated 576 sampling squares of 4.2 m grid size for a 1 ha field:

Then, if opponent 1 has the following grid layout which opponent 2 cannot see but has to identify from their 5 or 10 'guesstimates':

Hot spot no.	Diameter of hot spot/m	x coordinate(s)	y coordinate(s)
Hot spot 1	5	3	3
Hot spot 2	5	5	5
Hot spot 3	10	9–10	9–10
Hot spot 4	20	13–16	12–15
Hot spot 5	30	19–24	17–22

while opponent 2 has the following grid layout which opponent 1 cannot see but has to identify from their 5 or 10 'guesstimates':

Hot spot no.	Diameter of hot spot/m	x coordinate(s)	y coordinate(s)
Hot spot 1	5	23	2
Hot spot 2	5	22	2
Hot spot 3	10	20–21	3–4
Hot spot 4	20	16–19	5–8
Hot spot 5	30	10–15	1–6

a game allowing only five guesstimate coordinates per opponent's turn will last longer than one with 10. The full game will be based upon giving your opponent just the size of the field to be sampled (in hectares or metres) together with the number and sizes of the contaminated hot spots (in metres), while challenging them to pick out with 95% confidence the minimum number of tries to identify the positions of all the hot spots in the field given. Remember, you will know the required grid size and positions (as x–y coordinates) of all the hot spots for your opponent to guess because you are placing them—your opponent does not!

Sample preparation

Learning outcomes

Once you have read through this chapter, worked through the activities and problems, and reviewed the feedback, you should be able to:

1. understand the requirements for sample preparation for solid, liquid, and gaseous samples;

2. state the details of a range of equipment and methodologies used for sample preparation;

3. develop appropriate sample preparation strategies for samples with varying requirements.

3.1 Introduction

Before beginning sample preparation the analyst first has to identify the necessary procedures to undertake and then perform them so that the analyte, or property, can be readily measured using the chosen method of analysis. This step can easily be the most costly in terms of time and money and is the reason why much effort has been expended on methods of direct analysis, with the minimum of sample preparation. As we shall see, in this and later chapters, the processes of sample preparation and measurement must be chosen very carefully if you wish to answer correctly the questions posed in Chapter 1.

3.2 Factors to consider prior to sample preparation

Having identified the answers to our 'what' and 'why' questions that we posed earlier (i.e. what are we sampling/measuring and why are we performing this analysis) we are presented with a number of choices, as shown in Figure 3.1.

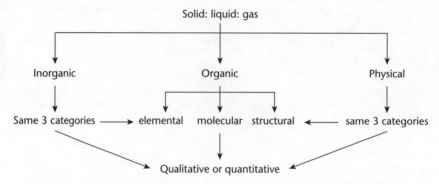

Figure 3.1 Analytical choices flowchart

The choices presented can be written as:

1. Is the sample a solid or liquid or gas, or a combination of phases?

2. For each of the above sample types are we considering a measurement of an inorganic or organic analyte, a physical property, or a combination?

3. Can each of the above be further specified as elemental, molecular, or structural measurements?

4. Are the measurements qualitative or quantitative in nature?

Before considering the answer to these questions, let us return to the problem of the exotic fish tank which we encountered in Chapter 1. There was some debate as to exactly what should be measured in any samples taken from the tank, with the key questions summarized as:

• What is our source or bulk starting material?

• Do we wish to measure the levels of nitrate or nitrite, or perhaps ammonium present, or even the total nitrogen content?

• Is the water in the tank clear or coloured?

• Is there any evidence of very fine solid material in suspension throughout the tank or of settled material at, or close to the bottom, in with the sediment?

• What other factors may affect the environment in the tank, for example, oxygenation, temperature?

We can propose the following possible scenarios:

1. A liquid (water-based) sample requiring a qualitative and quantitative, inorganic chemistry measurement of the molecular ions (i.e. nitrate, nitrite, and ammonium) in solution from the tank.

2. A liquid (water-based) sample requiring a qualitative and quantitative organic chemistry measurement of the molecular compounds (i.e. organic nitrogen-containing compounds) in solution from the tank.

3. A solid sample requiring a qualitative and quantitative inorganic chemistry measurement, of a mixed inorganic and organic material, of an element (i.e. the total nitrogen) to be measured in the sediment from the tank which is contaminated with various organic components (food, fish faeces, etc.).

It is important to note that in our above examples, the sample's matrix is also identified in terms of its 'type'. That is, is it inorganic, organic or a combination, in its make-up. As we will see, knowing the chemical make-up of the matrix as well, will help us to formulate how an analyte can be released from the sample itself, when this is required. Some examples of types of analyte and matrix-combinations were covered in column 1 of Table 2.2.

ACTIVITY 3.1 More possible scenarios

In the same way as for the 'fish tank' problem, identify, using the analytical choices flowchart, Figure 3.1, the possible scenarios for the following:

(a) a pesticide, dichlorvos, in a food crop which must not be above the maximum allowable concentration [Hint: The dichlorvos structure, physical and chemical properties may be found at the link: https://pubchem.ncbi.nlm.nih.gov/compound/dichlorvos];

(b) the levels of phosphorus in a soil sample from a field in order to calculate how much fertilizer to add. [Hint: The useful 'phosphorus' is often present, mainly as phosphate (PO_4^{3-}) in soil.]

The conclusion may be that we need to take samples which contain all three phases, solid, liquid, and gas. We may need to identify what is there (**qualitative** information), how much is there (**quantitative** information), and also to measure selected physical properties, for example, temperature and dissolved oxygen, as well as specific analytes of interest. So, we will have to consider carefully what is important to measure and how to ensure that any measurement made is truly representative of the original bulk material in the tank. This is not always an easy task and is one that can be both demanding and exacting. For this reason, a sound analytical plan should be agreed upon that limits the errors throughout.

It is worth noting the terms '**sample processing**' and '**sample preparation**' are both used to describe procedures performed on the sample. These may well overlap with some of the latter parts of the sampling step discussed in Chapter 2, such that anything that is done to the sample to preserve or select the analyte or property can be considered as sample processing.

Assuming that a suitable, **representative sample** has been acquired, as discussed in Chapter 2, the next step is to decide on the procedures necessary to bring the sample safely[1] into a state where the analyte (or its property) is in a stable form suitable for the chosen measurement technique to be employed. Whatever its form and composition, the representative sample must be processed so that what is to be measured in the source material is not lost or changed, that is, it remains identical to that in the original, or bulk, material, throughout. This is an important consideration because many measurement techniques require the sample to be in a particular form, often the liquid state, so some form of sample preparation will be required if the original sample is a solid or gas, or mixed phase. That said, both solid and gaseous sample introduction systems are also known for certain measurement techniques. So, the analyte (property) to be measured in a sample, the sample preparation procedure, and chosen measurement technique are linked together, such that any preparative step must, by the end of the procedure, render the sample in such a state that the analyte (property) can be measured at the level **finally** present in the processed sample.

There are several further considerations that may be important at this stage:

Separation. The sample preparation step should identify whether the analyte is to be separated from the **matrix** of the sample by a suitable **separation technique**. This is often performed in order to improve the **selectivity** for the analyte with the chosen measurement technique, and to reduce **interference** effects that might be encountered later on. In Chapter 4 we will see

---

[1] All procedures described should be performed in a safe manner in accordance with local health and safety regulations and procedures, with full cognizance of any chemical hazards and measures put in place to minimize risks.

that elimination of interferences can often be achieved during the measurement step, thereby simplifying the preparation. In addition, do liquid or gaseous samples require any filtering step to separate out or capture the desired phase(s)? Does the sample require any digestion or extraction procedures to be employed in order to release the analyte and how is this achieved?

Pre-concentration. In addition to any separation process, it will be necessary to identify whether pre-concentration of the analyte should be undertaken. This is in order to bring the levels of analyte within the **sensitivity** and **limit of quantitation (LOQ)** of the chosen measurement technique (see Chapter 5 and the glossary for these terms).

Physical state. In addition to the main physical state of the sample (solid, liquid, or gas), is a solid or gaseous sample wet or dry? If it is wet due to water or another volatile constituent such as an organic solvent then is this important to your measurement? If not, how should you dry the samples? For solid samples, is any coarse sieving or riffling required before sub-sampling, in either the wet or the dry states, and to what size cut-off? Is it necessary to grind the sample to reduce the particle size of the samples collected and how can this be achieved? All these have to be considered, so that a representative sub-sample may be taken from the bulk.

Size. In taking suitable sub-samples, how much (mass or volume) and how many test samples are to be taken? Remember also that when the prepared sample is made up to a final volume, there will be a dilution factor which will determine the concentration range of the analyte.

Procedural blank. It will be necessary to correct for any contribution to the total amount of analyte resulting from the process or procedure undertaken on the sample. This can be considered to be a form of contamination by the process, one that is introduced through using other materials and reagents that come into contact with that sample. This correction can be made by producing **procedural blanks** which are taken through entirely the same process of sample preparation as the samples themselves, but contain no sample. The procedural blanks are then run alongside the prepared samples in order to measure their contribution. For example, a river water sample with a lot of organic material in solution was heated with known volumes of concentrated nitric acid and 30% hydrogen peroxide to destroy the organic compounds present and remove any interference. An ultra-pure water sample of the same volume as the sample was treated identically and used as the procedural blank. When later measured, the nickel in the sample was found to be $35\,\mu g\,L^{-1}$ while that of the procedural blank gave a value of $10\,\mu g\,L^{-1}$. The actual value in the water sample when corrected is $25\,\mu g\,L^{-1}$. Procedural blanks are particularly important when measuring low concentrations in processed samples.

Reference materials. Wherever possible a reference material should be analysed alongside the sample to **validate** the method of analysis (see Chapter 9). Validation refers to the process of establishing the accuracy of the method undertaken. A variety of different reference materials are available, and are described in detail in Chapter 6.

We can then begin to consider the various preparative techniques which will bring the sample into a form that will allow the measurement to be made.

3.3 Solid samples

In many cases, solids will require more treatment than liquids and gases. The fact that the analyte may be variably distributed throughout the matrix, as well as being effectively 'locked' in it means that sample preparation is an important step. Key requirements are:

- taking a representative sample and sub-samples;
- homogenization of the analyte or the property.

Before we go any further, it is worth mentioning the concept of **direct solids analysis**. For a number of sample types, and for certain measurements, it is possible to use minimal sample preparation of grinding, sieving and in some cases just surface polishing, in order to just present the solid to the measurement technique. The advantages of shorter time and lower cost for sample preparation are important drivers for this approach. We will be covering some examples of those techniques in Chapter 4, but for the moment we will assume that a variety of **sample treatments** (often ambiguously referred to as pre-treatments) may be necessary to release the analyte and present it in a form ready for measurement. These are described in the following sections.

3.3.1 Drying

We tend to associate 'wet' and 'dry' samples with the amount of water being present. In our water-rich world this is understandable; however, samples may have other liquid solvents present, such as organic solvents. Depending on conditions, the solvent may be volatile, reactive, stable, or inert; so the sample may change in composition with time as conditions change. If the solvent is important to the measurement this will need to be taken into account. Two questions need to be addressed before a method of drying is decided upon:

- Do I need to know what, and how much, liquid solvent is present?
- Will I lose or modify my analyte (or the property) to be measured in my sample?

To measure the amount of solvent a simple weighing process before and after the drying stage may be all that is required. If you need to identify the solvent present, then a suitable collection or trapping technique should be employed. If chosen correctly both qualitative and quantitative measurements can be achieved in one step. Some analytes are themselves volatile, for example, pesticides, some mercury compounds, and gases dissolved in liquids. A number of drying techniques which address these concerns are listed in Table 3.1.

ACTIVITY 3.2 Drying techniques

Identify what drying techniques can be used to prepare the samples obtained in Activity 3.1 [Hint: Consider the thermal properties of the analytes you identified in Activity 3.1, then the composition of the samples themselves and then consider Table 3.1.]:

(a) a pesticide, dichlorvos, in a food crop which must not be above the maximum allowable concentration;

(b) phosphorus in a soil sample from a field in order to calculate how much fertilizer to add.

3.3.2 Size reduction

We will see in this and later chapters that representative sampling and sub-sampling processes for particulate materials are dependent upon particle size; and if you can reduce the particle size of the original sample taken from the source material, then it is possible to take even smaller sub-samples that are representative of the original sample. These smaller laboratory sub-samples

Table 3.1 Sample drying techniques

Drying technique	Temperature range	Timescale	Sample types
Air drying	Ambient	Usually days to weeks—often slow removal of volatiles	Water removal from soils, sediments, leaves, etc.
Thermal or cabinet drying	As set from 30°C up to 250°C, but often at 105°C	Usually hours to days; accelerated removal of volatiles	Water removal from soils, sediments, leaves, etc.
Rotary evaporation	Ambient to 100°C	Usually minutes to hours	Organic solvents and solutions of dissolved organic preparations
Rotary evaporation with vacuum	Ambient to 100°C; reduced pressure to aid loss of volatiles	Usually minutes to hours	Organic solvents and solutions of dissolved organic preparations
Thermal cabinet with vacuum	As set from 30°C up to 250°C; reduced pressure to aid loss of volatiles	Usually hours to days; for more difficult-to-remove volatiles	Water and some higher temperature organic solvent removal from soils, sediments, leaves, etc.
Freeze drying (under vacuum)	−40 to −80°C; very low pressure (high vacuum) conditions	Usually days (~24 to 72 hours) but can be less; dependent upon sample type; preferred technique for flora and fauna	Organic samples, including plant material and animal tissues; to preserve sample with minimum loss of integrity; also used for soils, sediments, etc. when certain volatiles suspected
Muffle furnace	As set from 250°C up to 1000°C	From minutes to hours	Usually inorganic high temperature materials to ensure loss of volatiles and desorption of contaminants

are often known as **test samples**. There are three further consequences of performing this size reduction process:

1. It will modify some of the physical properties of the original sample. So **some** physical and physico-chemical properties must be measured before this process is undertaken. Examples include the particle size distribution, bulk density, surface area and related activity properties, crystallinity, and sensitive crystalline states of the material.

2. The process of size reduction normally involves energy being put into the sample to break it down, with a consequent increase in temperature. This can result in losses of volatiles and changes to the concentration of labile analytes (i.e. degradation or denaturing of sensitive components) unless steps are taken to reduce this.

3. The decrease in particle size and increase in particle number for the same mass of material results in a dramatic increase in surface area available for reactions to occur. Therefore, dissolution and extraction procedures to get at the analyte of interest are far more rapid and efficient.

The way in which the process of size reduction works in our favour is shown by the simple examples below:

EXAMPLE 3.1

More, smaller particles are better than fewer, larger particles
This allows a smaller test sample weight to be taken that is representative. If we assume that all the particles in our solid sample are based upon spheres, then we know that each particle's volume $= \frac{4}{3}\pi r^3$.

If $r = 0.5$ mm then the volume of 1 particle $= 0.524$ mm^3

If $r = 0.05$ mm then the volume of 1000 particles $= 0.524$ mm^3

Hence, for a 10× reduction in particle size we have increased the number of particles by 1000 in the same sample volume, so we can take a much smaller sample volume (or weight) and still have a representative sample.

EXAMPLE 3.2

Smaller particles have a higher surface area
By reducing the particle size you expose the same amount of material to a reaction but with an increased surface area to react. If each particle's surface area is given by $4\pi r^2$ then:

If $r = 0.5$ mm then the surface area of each particle $= 3.14$ mm^2

If $r = 0.05$ mm then the surface area of each particle $= 3.14 \times 10^{-2}$ mm^2

Hence, for a 10× reduction in particle size, the combined particle surface area for the same mass of material increases by 100. This presents a material which will digest and extract more efficiently and on a shorter time scale.

There are a variety of particle size reduction techniques and a detailed description is beyond the scope of this chapter. However, some of the more important sample size reduction procedures are shown in Table 3.2.

3.3.2.1 Contamination

It was noted earlier that the reduction of a solid sample's particle size requires energy. This can result in losses or changes to the material. While temperature effects can be reduced by cooling, the associated problem of contamination from the grinding medium must be considered. When the measure of hardness (MOH) of the sample is close to, or greater than, that of the grinding medium, then contamination can be excessive. For example quartz, a silica based material, has a MOH of ~7 (Table 3.3), so one would obviously avoid grinding a sample that contained appreciable quantities of quartz or a silica-based material using an agate pestle and mortar; particularly if you wished to measure the silicon content! The relative difference in

Table 3.2 Size reduction techniques

Size reduction technique	Comments	Further notes
Pestle and mortar	The classical grinding technique of the laboratory	Materials from which they are made include glass, agate, porcelain, and metals
Percussion mortar (see Figure 3.2)	Often used in the sub-sampling and initial size reduction of hard, brittle materials	May include size reduction of carbides, nitrides, borides, cement, clay, shale, mica, bone, and even some highly crystalline metals/metalloids
Ball mill	Classical roller mill where material is contained within a cylindrical container together with grinding elements (usually ball-shaped)	Contamination is dependent upon the relative difference in the MOH (see Section 3.3.2.1); balls may be made of corundum or steel or similar hard materials; very effective size reduction technique
Cryogenic mill	To reduce the temperature and avoid losses due to volatility or chemical changes	Can also be used to make soft materials brittle, allowing fragmentation and particle size reduction, e.g. plastics, organic materials, etc.
Ring and puck	This is used with a vibrating stage that allows very efficient size reduction in a short time	Materials from which it is made include steel, carborundum, and tungsten carbide
Roller or trundle mill	Particulate material is placed inside a cylindrical container that is 'rolled' or 'trundled'. The sample grinds against itself to just break up aggregates and slowly reduce its primary particle size	Less contamination but only reduces to its primary particle size, with time; can use plastic bottles in some cases
Hammer mill	Often used with coarse material for primary size reduction step	Usually metallic in construction; made from steel or similar hardened material
Disc mill and pin mill	Former based upon rotating discs that are close together and latter on rotating discs with pins set in them through which particulate material is poured	Usually metallic in construction; made from steel or similar hardened material

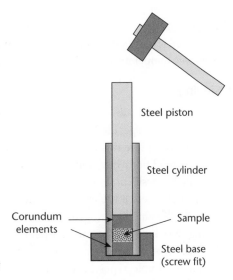

Figure 3.2 Example of a percussion mortar

MOH between the material to be ground and that of the grinding medium will often govern the degree of contamination.

So, one has to consider that grinding will contribute some contamination and that this will increase with grinding time. Contamination can be kept to a minimum by careful selection of the grinding technique, grinding medium, and period of grinding.

A related problem is encountered when using drilling to produce metal shavings that will both be representative of the bulk and allow for easier digestion. Specialist drill bits, of suitable MOH, can be acquired to reduce contamination in these cases. It is noted that i) dried plant material, while mainly having an organic make-up which is low on the MOH scale, <3, can also

Table 3.3 Measure of hardness (MOH)—extended scale

Hardness	Typical materials
1	Talc
2	Gypsum
2.5	Fingernail, pure gold, silver, aluminium
3	Calcite, copper coin
4	Fluorite
4.5	Platinum, iron
5	Apatite
6	Orthoclase, titanium, spectrolite
6.5	Steel file, iron pyrite, glass, vitreous silica
7	Quartz, amethyst, citrine, agate, porcelain (7+)
7.5	Garnet
8	Hardened steel, zirconia, topaz, beryl, emerald, aquamarine
9	Corundum, tungsten carbide, ruby, sapphire
9.5	Carborundum, silicon carbide
10	Diamond
>10	Aggregated diamond nanorods

contain inorganic minerals (e.g. silica / SiO_2) in lower concentration but higher on the scale; and ii) very soft materials (e.g. certain polymers, plastics and organic materials) may require freezing / very low temperatures, before they will grind.

3.3.2.2 Sieving

After samples have been subjected to a size reduction, it is often necessary to collect sub-samples below a certain specified size, particularly as part of a representative sampling step. The simplest way to achieve this is to sieve the ground sample. Sieving techniques are varied and can be performed both 'dry' and 'wet', using vibrating stages or by hand. The sieves can be made of a metal mesh (brass or steel or even electro-formed nickel) and even of plastic (polypropylene, nylon, etc.); the latter to reduce possible metallic contamination where this may be necessary. Sieving below ~100 µm can be difficult if the material is dry because it causes blinding (blocking) of the mesh. Therefore, the quantity of material to be sieved, the sieve mesh (hole size), and the diameter of the sieve itself have to be carefully considered (Table 3.4). Many sample types have to be ground prior to a digestion or an extraction process, in order to release the analyte of interest. Such examples can include soils or freeze-dried biological materials. It is not unusual in these cases for particles below 250 µm or even 180 µm size to be collected.

ACTIVITY 3.3 Grinding and sieving techniques

Identify what grinding and sieving techniques can be used to prepare the samples obtained in Activity 3.2 [Hint: Consider the physical properties of the samples themselves (made of, MOH etc?) first and then consider Tables 3.2, 3.3 and 3.4 together with section 3.3.2.]:

(a) a pesticide, dichlorvos, in a food crop which must not be above the maximum allowable concentration;

(b) phosphorus in a soil sample from a field in order to calculate how much fertilizer to add.

Standard mesh sizes for sieves are given in Table 3.4. It is worth noting that sieve sizes do not take particle morphology into account, that is, the shape of the particle. Hence, the smallest dimension of a particle, ground or otherwise will dictate its passage through a sieve.

3.3.3 Dissolution

Dissolution, or **digestion**, of a solid sample is the process of getting the sample, or a component of it such as the analyte, efficiently into solution. There are a number of approaches that you can take, often tailored to the type of sample, the analyte or the technique used to make the measurement. The overall decision-making process is represented in Figure 3.3.

3.3.3.1 Simple solvent dissolution

This simplest case is where a suitable solvent can used to dissolve the sample such that the analyte is entirely in solution. The simplest case would be the analysis of a water-soluble material where the analyte is released into solution. The application of heat may also aid this dissolution process. Such a happy state of affairs should be considered before any other more involved and complex approach is taken!

Table 3.4 Sieve mesh sizes

Aperture size		Mesh no.	
μm	mm	UK	USA
8000	8.000	n/a	5/16 in
6700	6.700	1	0.265 in
5600	5.600	3	3.5
4750	4.750	3.5	n/a
4000	4.000	4	5
3350	3.350	5	6
2800	2.800	6	7
2360	2.360	7	8
2000	2.000	8	10
1700	1.700	10	12
1400	1.400	12	14
1180	1.180	14	16
1000	1.000	16	18
850	0.850	18	20
710	0.710	22	25
600	0.600	25	30
500	0.500	30	35
425	0.425	36	40
355	0.355	44	45
300	0.300	52	50
250	0.250	60	60
212	0.212	72	70
180	0.180	85	80
150	0.150	100	100
125	0.125	120	120
106	0.106	150	140
90	0.090	170	170
75	0.075	200	200
63	0.063	240	230
53	0.053	300	270
45	0.045	350	325
38	0.038	400	400
32	0.032	440	450
25	0.025	n/a	500
20	0.020	n/a	635

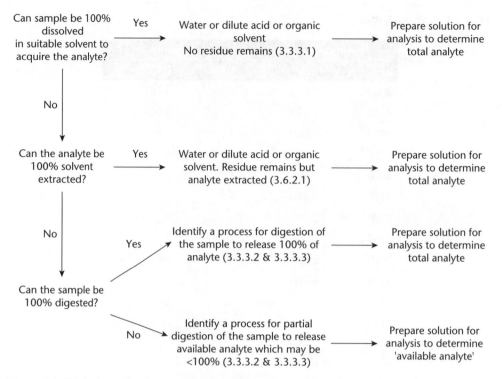

Figure 3.3 Decision-making process for dissolution of a sample with relevant sections of the text given in brackets

The process where the analyte is entirely in solution but some solid is left undissolved, is best considered under **solvent extraction**, covered in more detail in Section 3.6.

3.3.3.2 Dry ashing

Organic materials are generally decomposed by relatively low temperatures. Carbonaceous (carbon-based) compounds, held in a suitable crucible, can be converted to CO_2 and H_2O by gentle heating within a furnace. Temperatures between 300 and 450°C are often employed in air and various pure compounds may be added and mixed in with the solid (particularly if the sample is ground) to aid oxidation of the organic matter. The inorganic ash left would normally be the stable compounds (e.g. oxides) of the remaining elements, which result at the temperature range used. This inorganic ash may be dissolved in a suitable acid afterwards knowing that there is no organic matter left to deal with. Use of this method is dependent upon the volatility of the element-containing compounds produced during the ashing process. Unfortunately, if the conditions are not controlled, loss of the analyte(s) can occur. Elements such as As, Cd, Pb, Hg, and S, among others, can be easily lost at the temperatures encountered because of the formation of halides or volatile oxides (e.g. SO_2) from the sample matrix and even the reduction to their elemental state (intermediate carbon-reduction process). The application of suitable acids and other agents at the beginning of the ashing process to the organic sample matter may reduce these effects. However, a knowledge of the chemistry and properties of the components present is usually required before employing this technique.

3.3.3.3 Digestion

Digestion is the process whereby a sample is broken down chemically, often with the application of heat, to its component elements. As a consequence, it is used to prepare samples for elemental analysis, because the molecular structure is destroyed. The term digestion is also often used in the context of more gentle chemical treatments, but we shall define it here as meaning the complete breakdown of the sample into its constituent elements, or simple anions and cations in solution. Digestion methods can be divided into two main classes, **open vessel** and **closed vessel**.

Open vessel digestion

Digestion is usually performed in a beaker, flask, digestion tube, or crucible, usually with the application of heat, within a fume cupboard. The vessel may be covered (e.g. with a watch glass, lid, etc.) but the reaction is performed at atmospheric pressure. Heat may be supplied by a hot plate or hot block, a Bunsen or Meker burner, and even a muffle furnace. The reaction is always performed under controlled conditions for health and safety reasons, and the chemistry involved is chosen to reflect this. The quantity of sample, chemistry, and speed of reaction must be carefully considered. For example, oxidizing agents and organic matter can form explosive mixtures, while the addition of concentrated acids to materials that evolve gases in their reaction may be difficult to control. One example of the open vessel procedure would be the use of a **Kjeldahl-type flask** in a **hot-block** heater, which employs a mixture of concentrated sulfuric and nitric acids, for the **wet digestion** of an organic solid sample (e.g. a plastic) in order to release elements into solution for measurement.

Closed vessel digestion

Digestion is performed in a contained or sealed system. This allows the digestion mixture (e.g. aqueous solutions of acids, alkalis, oxidizing agents, or mixtures of organic liquids) to attain higher pressures and temperatures, resulting in increased rates of reaction, greater efficiencies of extraction, and with no loss of volatile analytes that might be the case for an open vessel digestion. Typical closed vessels include **digestion bombs**.[2] The energy source can be microwave energy (e.g. when using screw-top PTFE vessel with a pressure release option), a heating oven or muffle furnace (e.g. using metal-casing containers with various liners of different resistant materials), or just the exothermic reaction from reagents that are added to the sealed system (e.g. oxygen bomb with O_2, Parr bomb using Na_2O_2 for fusion melts, etc.). Microwave bombs are considered to be medium pressure vessels while the other examples are considered to be high pressure. The modern laboratory will usually have programmable microwave instruments available, together with a range of different bomb options, for both digestion and extraction procedures to be employed. One example of a closed vessel reaction would be the digestion of a mainly organic sample within a microwave bomb, using a mixture of concentrated nitric acid and hydrogen peroxide, for the determination of Cd, Hg, and As. Here the mainly organic material could be a paint sample, scraped from a wall in order to look for levels of banned pigments. The sample could be cryogenically ground and then sieved in order to prepare for its efficient digestion within the microwave bomb.

Wet digestion

In many cases solid samples will require a more involved process to acquire the analyte of interest in solution. A common procedure is the use of corrosive materials such as acids and alkalis to **digest** the sample. The process radically alters the sample by attacking the matrix chemically, destroying in full or in part the structure of the material. This results in the sample's components being in solution, hopefully including the analyte of interest! However,

---

[2] As the name implies, these are potentially explosive devices because the chemical reactions can release a considerable amount of gas. Strict safety procedures should be followed; a pressure release system is mandatory for microwave digestion.

the chemical attack on the sample should be tailored to release the analyte in its required form. Hence, harsh digestion techniques are often employed where an elemental measurement is to be employed. The degree of chemical attack is often governed by the type of digestant involved, its concentration, the sample matrix, and the temperature used. One simple 'rule of thumb' uses the concept of neutralization or opposition chemistry to decide on the required chemistry for the digestion. For example, where solid samples may have acidic and/or basic components with oxidizing and/or reducing properties, these can be attacked by their 'chemical opposite'. One example to illustrate these points might be the digestion of a sample of an important manganese ore, pyrolusite or MnO_2. It has oxidizing properties (with Mn in the IV state) and contains a metal oxide that has some basic character, particularly when reduced (MnO and Mn_2O_3 are basic oxides). It reacts with concentrated hydrochloric acid (HCl) with the evolution of chlorine gas (Cl_2), where the chloride is oxidized by the Mn IV, to produce the reduced state of Mn^{2+} in solution, as its chloride salt.

$$MnO_2 + 4\,HCl \rightarrow Mn^{2+} + 2\,Cl^- + Cl_2 + 2\,H_2O$$

One would expect all the MnO_2 (pyrolusite) to be digested and the manganese can then be measured. However, any associated silicate-based rock mineral with the manganese ore would not be dissolved and some residue may be left. If this residue contains no manganese then we have 100% of the analyte required from the sample but not 100% digestion of the matrix!

In some cases an analyte will be acquired by digesting the solid in a specific digestant that you know will not dissolve the whole matrix. This is called the **available** analyte and is seen when soil samples are attacked by different reagents of variable strengths in order to estimate such things as **plant availability** for certain components within the soil. The ability to measure these fractionation differences, in order to estimate what is available on biological timescales compared with a geological timescale, is becoming more important.

The range of digestion reagents most likely encountered together with their chemistries and usage are given in Table 3.5. Care must be exercised in all cases and the use of fume cupboards is expected. Oxidizing agents and the use of nitric acid with organic materials must be carefully assessed in terms of safety and risk of explosions.

Fusions

Certain materials can be particularly resistant to the general digestion techniques identified previously. Synthetic polymers known as plastics are often designed to have these qualities of resistance to corrosion, chemical attack, and weathering. Other refractory materials like the carbides of tungsten, boron, and silicon also have high MOH as well. When you add in materials such as the silicates, aluminosilicates, and the other associated naturally-occurring rocks and minerals then the 'difficult' list begins to get quite long. As shown in Table 3.5, the latter group of resistant materials based upon 'silicates' are only partly attacked by acids except for perhaps hydrofluoric acid, and only then when mixed with other acids.

In order to meet this challenge, a group of compounds known as **fluxes** can be used to prepare a **fusion** sample. Fluxes have the advantage of being strongly alkaline or acidic and even oxidizing when in their molten state, so are able to attack resistant samples at high temperatures. For example, the SiO_2 component of rocks and minerals is considered to be an acidic-like material (the oxide of a non-metal), hence alkaline fusion compounds like sodium carbonate, lithium carbonate, or lithium metaborate can digest these sample types when in their molten state. Table 3.6 shows some common fluxes and their properties.

Table 3.5 Digestion reagents

Digestion reagent	Chemical reactivity	Sample types	Comments
Nitric acid	HNO_3, very oxidizing when concentrated (69%); converts simple organic material into CO_2 and H_2O and evolves NO/NO_2; releases **elements** as soluble nitrates except for B, Al, Cr, Ti, Zr, PGMs, Au, (Sn)	Many metals, biological and organic materials; partial digestion of soils, sediments, ores. Simple digestions can be with 20–25% aqueous nitric acid, e.g. limestone, carbonates, etc.	Not generally good for plastics, aluminosilicates; destroys molecular systems
Nitric acid/ hydrogen peroxide	$HNO_3 + H_2O_2$, often 15–20% peroxide, increases the oxidative power of concentrated nitric acid (additional production of O_2)	For biological and organic materials; partial digestion of soils, sediments, ores to release elements and destroy organic matter	Efficient breakdown of organic matter; as above
Hydrochloric acid	HCl is not an oxidizing acid but can itself be oxidized in reactions; forms chloride salts with many metals, but Ag, Ti, Hg chlorides are insoluble	Metals including Fe and its alloys; partial digestion of some ores and minerals to release elements	Not for dissolving oxides of Al, Be, Cr, Sb, Sn, Si, Ti, Zr
Aqua regia	Oxidizing system comprising conc. HNO_3 + conc. HCl, usually in the ratio ~1:3; name given because it will dissolve gold. Based upon two reactions: $HNO_3(aq) + 3\ HCl(aq) \rightarrow NOCl(g) + Cl_2(g) + 2\ H_2O(l)$ and $2\ NOCl(g) \rightarrow 2\ NO(g) + Cl_2(g)$	Used for the digestion of precious metals (Pt, Rh, Au, etc.), some resistant sulfide minerals. For the partial digestion of soils, sediments, ores to release elements	Always produce fresh when required; used to extract available metals from clays, sediments, soils, minerals
Sulfuric acid	A strong acid with oxidizing properties when concentrated, H_2SO_4 in conc. form will remove the chemical equivalent of H_2O from organic material, leaving carbon backbone of organic samples; can be used to dissolve some samples if the sulfate of the product is soluble; used for high temperature work (b.p. = 330–337°C); often used with other acids	Used for some ores and minerals to break down samples; some metals will dissolve (Fe etc.)	Sulfates of Ba, Sr, and Pb are very insoluble. Will not dissolve silica-containing materials

Table 3.5 (*Continued*)

Digestion reagent	Chemical reactivity	Sample types	Comments
Sulfuric acid then nitric acid	Often referred to as 'wet digestion' method; conc. H_2SO_4 performs as stated previously on organic containing samples leaving a black carbonaceous material when heated; when conc. HNO_3 is added dropwise after this step, the carbon is converted to CO_2	Digestion of resistant organic material and all plastics, to release elements.	Usually performed in specialist tubes or beakers (Kjeldahl-type)
Hydrofluoric acid	A non-oxidizing, highly complexing acid, HF is used to dissolve/digest siliceous material $SiO_2 + 6\,HF \rightarrow H_2SiF_6 + 2H_2O \rightarrow SiF_4\uparrow$ with heating; often employed with other acids (HNO_3, H_2SO_4, H_3PO_4, etc.)	Ores, minerals, soils, sediments, silicon-containing materials	An acid which requires specialist handling and safety precautions. Some fluorides are insoluble while others are very volatile, Bi, Si, As, Ge, Se, etc.
Alkaline digestion	Usually NaOH or KOH solutions of required concentration	Acidic-based materials such as acidic oxides of elements (non-metallic character)	Some elements in solutions of higher pH are highly insoluble
Persulfate (potassium, sodium, ammonium salts)	Persulfate or the $S_2O_8^{2-}$ ion is one of the stronger oxidizing agents, dependent upon pH; effective under acid and alkaline conditions, it is often used to deliberately destroy refractory organic material and liberate elements	Organic solid suspensions in water, complex high molecular weight refractory organic samples	Care must be exercised in the use of strong oxidizing agents

When cooled, the fused material can be dissolved in a suitable aqueous-based solvent, such as water, simple acids or alkalis. Some direct solids analysis techniques such as X-ray fluorescence spectrometry[3] can be employed to measure the elemental constituents directly within the solid 'bead' cast from the molten flux.

While the molten salts shown in Table 3.6 can be very effective in the digestion of difficult samples, there are a number of disadvantages to take into account. Relatively large quantities

[3] See explanatory note 1 on page 70 regarding use of 'spectroscopic' terminology.

Table 3.6 Common fluxes for fusion sample preparation

Flux	Melting point/°C	Possible use	Crucible material
Na_2CO_3	850–851 (without additives)	Alkaline flux for Si- and Al-containing samples; can add oxidizing agents like KNO_3 for sulfides	Pt
KOH and NaOH	380 (anhydrous— lower if not); 318	Strongly alkaline fluxes for silicates, SiC, and a range of selected mineral types	Ni, Zr, Au
$LiBO_2$	845–849	Alkaline flux for high silicate-containing materials including some minerals and some ceramics	Pt/Au
$Li_2B_4O_7$	915	Acidic flux for alumina-based minerals, mixed mineral carbonates of group I and II metals, ZrO_2, TiO_2, etc.; not advised for high SiO_2 samples	Pt/Au
1:4 $Li_2B_4O_7$:$LiBO_2$	832 (eutectic) Used at 900–950	For very wide range of aluminosilicate materials including soils	Pt/Au, graphite
Na_2O_2	~495 with decomposition; O_2 evolved	Strongly oxidizing alkaline flux, for refractory oxides, sulfides, and carbides, acid-insoluble metals; not for open use with organic material	Zr, Ni, Fe (avoid Pt)
$K_2S_2O_7$ Pyrosulfate	Up to 419; region is purity dependent	An acidic flux for refractory oxide-containing samples including metal oxides	Pt, Vycor, fused-silica, Pt
B_2O_3 (or produced via heating H_3BO_3)	580	Acidic flux to use when alkali metals are determined	Pt

of flux are often used (requiring a high flux to sample mass ratio) to bring about an efficient fusion. This causes a large amount of salt matrix to deal with prior to measurement, often requiring a significant dilution to compensate when a solution measurement is required. High temperatures are required in the preparation of fusions (mainly between 300 and 1000°C) and some losses may occur unless conditions are controlled. Contaminants from the flux reagent and from the crucible used can be particular problems and the choice of crucible must therefore be carefully made. In addition to those shown in the table, porcelain, glassy carbon, graphite, magnesia (MgO), and even sintered silica crucibles have been used but all are highly dependent upon the type of flux employed.

> ## ACTIVITY 3.4   Extraction/digestion techniques
>
> Identify what extraction/digestion techniques can be used on the samples prepared in Activity 3.3 [Hint: See Figure 3.3 and the guide sections included in this; together with those sections' relevant tables (3.5, 3.7 and 3.8) and the examples given.]:
>
> (a) a pesticide, dichlorvos, in a food crop which must not be above the maximum allowable concentration;
>
> (b) phosphorus in a soil sample from a field in order to calculate how much fertilizer to add.

3.4 Liquid samples

There are advantages to sampling liquids, particularly when one considers the homogeneous distribution of the analyte when in solution. When simple aqueous matrices are involved, simple preparative processes can be used. However, some organic liquids or suspensions may be more complicated to prepare, depending on the analyte and the final measurement technique to be employed.

3.4.1 Sample stabilization

In Chapter 2, we stressed the importance of sample preservation, between the point of sampling and the laboratory, to maintain sample integrity. Liquid samples should be collected in suitable containers which minimize contamination or losses. They can be frozen or cooled (<4°C) or have the pH adjusted to ensure the integrity of the analyte(s) or property to be measured. When particulate matter is present the particles are in equilibrium with the liquid matrix which suspends them, together with any analyte that is in solution or adsorbed upon the surface of the solid. Any change to the liquid's composition, ionic strength, or pH could affect this equilibrium or even slowly dissolve the solid component. Careful selection of the conditions is therefore required in order to help answer the 'what' and 'why' questions of the analysis.

3.4.2 Liquid digestion

Liquid samples may require a digestion stage as part of the preparation step. This will obviously be dependent upon the 'what' and 'why' questions for the analysis and in many cases would come down to whether an elemental measurement is required. There are different approaches for aqueous and for organic liquids but note that the digestion techniques shown in Table 3.5 and processes shown in Section 3.3 may also be considered for certain liquids.

3.4.3 Drying

The simplest and one of the oldest methods to remove water from non-polar solvents has been the use of finely ground anhydrous sodium sulfate. A suitable quantity, in excess, is added to the solvent and stirred/shaken and finally left for a period to complete the drying process. This material then requires separation from the solvent and subsequent disposal.

An alternative approach is to use commercially available drying disc separation membranes which can reduce the time involved in removing residual waters from non-polar organic solvents. It is also noted to work on emulsified samples. This technology is based upon hydrophobic non-polar organic solvent-permeable filters and a vacuum manifold.

3.5 Gaseous samples

The reactivity of many gases means that sample preparation is mainly confined to trapping and/or stabilization of the gas prior to measurement. This dictates the use of one of two possible approaches:

1. direct measurement at the source or on a collected, stabilized gas sample;
2. direct adsorption or collection process at the point of sampling prior to release for its later measurement.

Adsorption and collection processes for the sampling of gases were covered in Chapter 2 and their measurement will be considered in Chapter 4.

3.6 Separation and extraction

3.6.1 Separation techniques

Two separation techniques that are commonly encountered are **filtration** and **centrifugation**, both of which have advantages and disadvantages.

3.6.1.1 Filtration

Filtration is used to separate solid and liquid phases, to enable collection and further treatment of one or both phases. Filtration depends on the pore size of the filtering medium, which determines the extent of particle breakthrough. This in turn depends on what is considered to be in solution, a definition which has changed over the years but is currently set at 0.22 μm, with everything above this size being deemed to be 'suspended'. In order to filter finer and finer material the pore (mesh) size has to decrease, but this can lead to an increase in filtration-time because of particles blocking the pores in the filter. Therefore, the solid to liquid ratio, or **loading**, can be an important factor that must be controlled. There is also the possibility of contamination from the filter itself, dissolution of the filter by corrosive or strong solvent solutions, and adsorption losses from the sample onto the filter. All of these effects are dependent upon the material from which the filter is made and the nature of the solutions that are poured through it. Careful selection of filter-type and materials is therefore required.

3.6.1.2 Centrifugation

Centrifugation achieves the same outcome as filtration. There is the requirement to re-suspend the solid phase in a suitable solvent in order to wash the collected material, and to then centrifuge again. This may require several repeats in order to efficiently wash the solid; however, it can result in less contamination and fewer losses of analyte when used correctly.

Centrifugation can be quite sophisticated in its ability to separate and fractionate solid phases of varying density, for example, the separation of red blood cells from plasma prior to analysis of the plasma component.

3.6.2 Pre-concentration

For solid or liquid samples it may be necessary to perform a separation and extraction step in order to **pre-concentrate** the analyte and/or separate it from the matrix. This may be necessary for a variety of reasons:

- the analyte is at a concentration below the **limit of detection (LOD)** of the measurement technique (see Section 9.2.1);
- the measurement technique suffers from **interferences** caused by the sample matrix or other analytes (see Section 5.6.2);
- the measurement technique is incompatible with the sample matrix.

Separation and pre-concentration can be achieved by:

1. separating the analyte, retaining it, and disposing of the matrix;
2. separating the interferent from the matrix, disposing of it, and retaining the sample containing the analyte at close to its original concentration.

In performing the separation given in (1) above, it may also be possible to pre-concentrate the analyte. If pre-concentration is required after a type (2) separation, then an extra process stage will need to be performed. In both cases, the process depends on the physical and chemical properties of the matrix, **interferent**, and analyte. A range of methods which can be used to separate and pre-concentrate are given in Table 3.7.

Table 3.7 Separation and pre-concentration techniques

Separation method	Operates on	Example	Comments
Volatility I	Volatile analyte	Hydride generation of As, Se, S, N, or cold-vapour of Hg	Reducing and acidic conditions; may suffer interference from transition metals
Volatility II	Volatile matrix	Evaporation of H_2O; loss of SiF_4; organic solvent	Pre-concentrates analyte; dependent upon relative volatility of analyte
Liquid–liquid extraction I	Organic analyte	Solvent extract organic analyte between cyclohexane and acetonitrile or between water and DCM	Preferential partitioning of the analyte between two immiscible solvents, with much greater solubility in one compared to the other
Liquid–liquid extraction II	Organometallic complex of analyte	Extraction of metal–APDC complexes into organic solvent from aqueous digest	As in liquid–liquid extraction I

Table 3.7 (*Continued*)

Separation method	Operates on	Example	Comments
Co-precipitation	Analyte	Adsorption of transition metals onto fresh $Fe(OH)_3$ precipitate	Retention of metal analytes onto another metal precipitate with surface adsorption properties; re-dissolve precipitate in acid to release all analytes
Solid phase extraction (SPE) by ion exchange	Analyte or matrix	Cr^{3+} retained on a cation exchange resin; CrO_4^{2-} and $HAsO_4^{2-}$ on an anion exchange resin	Metal and non-metal ions retained on oppositely charged ion exchange resins; released by changing pH or using competitive species
Solid phase extraction (SPE) by chelation	Analyte or matrix	Co^{2+} retained by Chelex type $-N(CH_2COO^-)_2$ groups on resin	Metal ions retained by chelating agents bound to resins and onto modified solid phase surfaces; less dependent on ionic strength of solutions; dependent on stability constant of complex; released by changing pH or using competitive species
Solid phase extraction (SPE) by van der Waals forces	Analyte or matrix	PAHs extracted from waters onto C_{18}-bonded silica discs; eluted using DCM and ethyl ethanoate	Organic analytes adsorbed onto modified organic solid phase surfaces; polarity of adsorbent dependent on analyte to be retained; released from adsorbent based upon solvent with competitive solubility and / or polarity effect; a type 2 separation or matrix adsorption clean-up step can be achieved using fine particulate graphite / carbon
Thin layer chromatography (TLC)	Analyte	Separation of saturated fatty acids from mono- and polyunsaturated fatty acids on modified particulate silica-coated glass plates	Organic analytes separated from matrix by differences in partitioning between solvent phase and polar solid adsorbent phase

3.6.2.1 Solvent extraction

Certain samples allow the absorption and permeation of solvents through their solid structure, releasing selected components into this solution by a dissolution process while still leaving a solid matrix behind. The efficiency of this process is dependent upon the analyte, the sample type, its physical form, and the solvent used. This approach may be considered as a relatively soft physico-chemical extraction process and one that can be performed within open or closed vessels. The careful application of heat and controlled interaction between solvent and sample will improve both the speed and efficiency of the process. This can be achieved in a number of ways, for example, via hot plate or isomantle, simple reflux, or Soxhlet system, microwave assisted or under high pressure and temperature, as shown in Table 3.8.

One example of this general process would be the extraction of a molecular analyte from an organic, biological material. A more specific example is the extraction of a herbicide, for example, 2,4-D (2,4-dichlorophenoxyacetic acid), which has been sprayed to deter weeds around

Table 3.8 Solvent extraction methods

Extraction technique	Sample type/ solvent type	Open or closed vessel	Temperature	Notes
Simple permeation with shaking, stirring or sonication	Solid or liquid; organic or inorganic	Open, e.g. beaker, conical flask, plastic sample tubes	Solvent temp. altered by environment: can pre-heat or cool solvent	Variable temperature using hot plate, microwave, sonication
Simple reflux using condenser	Solid or liquid; organic or inorganic	Open: round bottomed flask	Solvent temp. altered by isomantle heating	Temperature can be controlled with time
Soxhlet	Solid or liquid; organic or inorganic	Open: round bottomed flask with Soxhlet tube	Solvent temp. altered by isomantle heating and solvent type	Efficient extraction: temp. can be better controlled with solvent over time
Separating funnel	Solid or liquid; organic or inorganic	Partially closed or open	Solvent temp. altered by environment	Can pre-heat or cool solvent
Microwave extraction tube	Solid or liquid; organic or inorganic	Closed	Solvent temp. and pressure changed using microwave energy	Pressure-rated sealed vessels
Accelerated solvent extraction (ASE)	Solid or liquid; organic or inorganic	Closed	Solvent temp. altered by environment: extraction cell placed in oven	Efficient extraction: can extract at temp. above boiling point of solvent by increasing pressure in cell
Supercritical fluid extraction (SFE)	Solid or liquid; organic or inorganic	Closed	Solvent phase changed from gas to liquid: temp. and pressure altered by environment	Efficient extraction: can extract at temp. above boiling point of solvent by increasing pressure in cell to liquefy gas used
Separating funnel	Liquid/organic or inorganic liquid+	Partially closed/open	Solvent temp. altered by environment	Variable temperature. Partitioning is governed by solubility or migration effects

a plant material grown for feeding cattle. This food-based plant should not have high levels of this pesticide in it. First, it would be representatively sampled, freeze-dried to help preserve the analyte, then carefully ground and sieved to present a fine material with high surface area to allow release of the analyte. Using a solvent such as methanol, the 2,4-D could then be extracted. The 2,4-D is highly soluble in methanol (>50 g per 100 mL) and when the ground sample and methanol are shaken together the organic solvent is able to permeate the organic plant matrix and efficiently extract the analyte, in its unchanged molecular form, into solution, suitable for later measurement. Solid plant material will still be left after the extraction but repeated washing and shaking of the solid with methanol will ensure a high efficiency of extraction. However, it is possible to avoid this labour-intensive approach by undertaking a Soxhlet extraction process. This employs both an elevated temperature and continuous interaction between pure solvent vapour and the sample, thereby resulting in a highly efficient extraction under controlled conditions. Using the rule of thumb that 'like dissolves like'; in this case a polar organic solvent is used for a polar organic analyte. In addition, like-materials are often compatible and here we have a slightly polar organic solid sample which can absorb a polar organic solvent.

3.6.2.2 Solid phase extraction

Solid phase extraction (SPE) using ion exchange, chelating, and organic coated substrates is widely used to selectively separate analytes from other components in a liquid sample, and also allow pre-concentration where necessary. As an example of separation and pre-concentration being performed on a processed sample, consider the problem of measuring the levels of morphine in blood plasma. To obtain the required sensitivity for such a measurement would necessitate the use of mass spectrometry, but the proteins and other organic components in the blood plasma would not be handled well by the instrument. Dilution of the plasma matrix to a handleable level would only lower the concentration of the analyte even further, below the limit of detection. Use of a C_{18} SPE column would allow the morphine to be retained on the surface if the pH is controlled using an appropriate buffer, while the other components in the plasma remain in solution. So 1.0 mL of plasma, whose pH has been adjusted to 9.5, is poured through the SPE cartridge which retains the analytes of interest, and the cartridge washed to remove the associated matrix. The morphine and other metabolites are then released by eluting with 0.2 mL of a methanol/acetonitrile/orthophosphoric acid mixture (polarity and solubility effects). The 5× increase in analyte concentration together with the removal of the plasma matrix will allow the instrument to easily detect and measure the morphine without interference.

It should be noted that some separation methods are analytical techniques in their own right, for example, high performance liquid **chromatography** (HPLC) and gas chromatography (GC). These instrument-based approaches to separation will be discussed in the following measurement section of Chapter 4, together with other instrumental techniques.

ACTIVITY 3.5 Separation/clean-up/pre-concentration techniques

Identify what separation/clean-up/pre-concentration techniques can be used on the samples prepared in Activity 3.4 [Hint: Consider the use of non-miscible solvents for the organic systems and / or the use of SPE for both (a) and (b). See tables 3.7 and 3.8.]:

(a) a pesticide, dichlorvos, in a food crop which must not be above the maximum allowable concentration;

(b) phosphorus in a soil sample from a field in order to calculate how much fertilizer to add.

PROBLEM 3.1

Fish tank problem

Let us once again consider our exotic fish case study. We had previously identified the following categories for the three different possible scenarios focussed around the measurement of nitrogen-containing samples from our fish tank:

1. A liquid, water-based sample requiring a qualitative and quantitative inorganic chemistry measurement, that is quantitative, of the molecular ions (i.e. nitrate, nitrite and ammonium) in solution from the tank.

2. A liquid, water-based sample requiring an organic chemistry measurement, that is quantitative, of molecular compounds; that is, organic nitrogen-containing compounds to be measured in solution from the tank.

3. A solid sample requiring an inorganic chemistry measurement, from a mixed inorganic and organic material that is quantitative, of an element; that is, the total nitrogen content to be measured in the sediment (mainly highly insoluble silicate and aluminosilicate mineral granules) from the tank which is contaminated with various organic components (fish food, fish faeces, dead organic matter, etc.).

For each of the three scenarios shown above, and based upon your experience of using the previous sections with figure 3.3 and tables 3.1 to 3.8, describe the sample preparation processes involved which will result in the required analyte being ready for measurement in solution.

Feedback on activities and problems

 FEEDBACK ON ACTIVITY 3.1

More possible scenarios

In the same way as for the 'fish tank' problem, identify the possible scenarios for the following:

(a) a pesticide, dichlorvos, in a food crop which must not be above the maximum allowable concentration;

We would require a qualitative and quantitative, organic chemistry measurement of a molecular compound (i.e. the pesticide dichlorvos) in a solid sample (the food crop; organic plant material). The quantitative measurement must be made at a level lower than the limit specified. This limit can be found within the link given for Activity 3.1, for various food crops or from an internet search technique using key words 'pesticide residue' and 'dichlorvos'.

(b) the levels of phosphorus in a soil sample from a field in order to calculate how much fertilizer to add.

We would require a qualitative and quantitative, inorganic chemistry measurement of phosphorus, but actually measured as the molecular ion, phosphate (PO_4^{3-}) in a mixed organic and inorganic solid sample (soil).

FEEDBACK ON ACTIVITY 3.2

Drying techniques

Identify what drying techniques can be used to prepare the samples obtained in Activity 3.1:

(a) *a pesticide, dichlorvos, in a food crop which must not be above the maximum allowable concentration;*

Dichlorvos is an organic compound which is volatile (see physical properties, as directed) and can be lost at elevated temperatures and over time at ambient temperatures. Thermal drying even at 40°C could result in losses, so freeze-drying may be appropriate to remove water from a food crop – an organic plant material; e.g. tomatoes, peppers etc. The alternative is to avoid drying and extract from the 'wet' sample directly.

(b) *phosphorus in a soil sample from a field in order to calculate how much fertilizer to add.*

Phosphate is a non-volatile inorganic compound so thermal drying at 105°C could be used to remove mainly water from a relatively non-volatile soil sample.

FEEDBACK ON ACTIVITY 3.3

Grinding and sieving techniques

Identify what grinding and sieving techniques can be used on the samples prepared in Activity 3.2:

(a) *a pesticide, dichlorvos, in a food crop which must not be above the maximum allowable concentration;*

The food crop would be a plant material which, after drying, could be ground, carefully, with a pestle and mortar of suitable MOH (glass, agate, porcelain etc.). If the crop contains edible and non-edible parts it may be necessary to separate these for individual treatment and analysis so that the level of dichlorvos in the edible part can be determined. The material should be ground with a pestle and mortar to reduce the particle size (<250 μm) so that a representative sample of the bulk can be taken (suitable metallic or plastic sieve) and to aid extraction of the analyte during sample preparation. The alternative, direct 'wet' sample approach would be to chop/shred the non-dried crop sample up into very small pieces ready for extraction. However, representative sub-sampling becomes a problem as larger quantities would be required.

(b) *phosphorus in a soil sample from a field in order to calculate how much fertilizer to add.*

The soil can be ground with a pestle and mortar or in a ball mill to break up the agglomerated grains so that they can pass through a sieve (<250 μm, plastic or metallic). This separates the more representative soil particles from the larger grains and other contaminants and aids analyte extraction during sample preparation. Depending on the type of soil (e.g. limestone, clay, sandy, etc.) it may be necessary to choose a particular type of grinding medium (of suitable MOH, e.g. porcelain etc.) in order to avoid contamination.

FEEDBACK ON ACTIVITY 3.4

Extraction/digestion techniques

Identify what extraction/digestion techniques can be used on the samples prepared in Activity 3.3:

(a) *a pesticide, dichlorvos, in a food crop which must not be above the maximum allowable concentration;*

It will be necessary to extract the dichlorvos from a known mass of the sample while maintaining the integrity of the analyte, so a relatively mild extraction procedure is required. Under these circumstances it is important that the particle size has been reduced sufficiently to release the analyte, or it can be extracted directly from the sample by an efficient solvent. Dichlorvos is an organic compound , so solvents such as dichloromethane, acetonitrile, methanol and ethyl acetate, etc., could be used for extraction. A lower to near room temperature extraction may be applicable. Using the reflux condenser soxhlet extraction technique with the more volatile organic solvents would maximize extraction efficiency while reducing losses. The alternative is to shake a suitable solvent directly with the known, chopped/shred 'wet' sample mass, and repeat this two more times on the same pre-extracted sample and combine all the extracts.

(b) *phosphorus in a soil sample from a field in order to calculate how much fertilizer to add.*

If phosphate must be determined then the extraction should be designed to extract the phosphate ion only, from a known mass of the sample. However, this is not always easy to achieve. Simple acid digestion techniques (using nitric and/or hydrochloric acids) of dried, ground and sieved soil samples are possible for releasing phosphate, but there will be some variation in extraction efficiency. Ideally, one would be after the 'plant-available phosphorus'. So, many extraction methods from soils for phosphorus are **operationally defined**; that is, the fraction or **species** of phosphorus extracted is determined by the method. For example, the Bray II[4] method for soil phosphorus requires extraction into a mixture of NH_4F, HCl, CH_3COOH, and NH_4NO_3.

 FEEDBACK ON ACTIVITY 3.5

Separation/clean-up/pre-concentration techniques

Identify what separation/clean-up/pre-concentration techniques can be used on the samples prepared in Activity 3.4:

(a) *a pesticide, dichlorvos, in a food crop which must not be above the maximum allowable concentration;*

If dichlorvos is extracted into an organic solvent then this can be further separated from co-extracted matrix components by liquid–liquid extraction using two immiscible solvents, or solid phase extraction (SPE) using a C_{18} substrate. In the latter case, the dichlorvos would be retained by the C_{18} then subsequently eluted in a stronger solvent. Pre-concentration can be achieved in both cases by limiting the volume of the final extracting solvent. Alternatively, a type 2 matrix removal/clean up may be achieved after the original solvent extraction step using fine particulate carbon to absorb matrix components, leaving the analyte in the original solvent system; then filtering or centrifuging.

---

[4] Wuenscher, R., Unterfrauner, H., Peticzka, R., and Zehetner, F. 2015. A comparison of 14 soil phosphorus extraction methods applied to 50 agricultural soils from Central Europe. *Plant, Soil and Environment, 61*(2), pp.86–96.

(b) phosphorus in a soil sample from a field in order to calculate how much fertilizer to add.

If phosphorus is extracted in an anionic form (i.e. as phosphate; PO_4^{3-}) then sample clean-up and pre-concentration can be achieved using SPE onto an anion exchange column. The anionic phosphate is retained on the column, the matrix is washed away, and then the phosphate is eluted from the column using a smaller volume of solvent.

FEEDBACK ON PROBLEM 3.1

Fish tank problem

Let us once again consider our exotic fish case study. We had previously identified the following categories for the three different possible scenarios focussed around the measurement of nitrogen-containing samples from our fish tank:

1. *A liquid, water-based sample requiring a qualitative and quantitative inorganic chemistry measurement of the molecular ions (i.e. nitrate, nitrite and ammonium) in solution from the tank.*

Being a water sample from a fish tank, we would need to consider processing and stabilizing our sample as soon as possible. The possibility of losses because of the presence of particulate matter absorbing, or biological activity using up, our nitrogen-containing analytes in the sample is high; the integrity of the sample and the target analytes of interest would be lost. Freezing of the sample as soon as possible after collection may be considered if immediate processing is not readily available and would allow a delay. To isolate those analytes in solution from particulate matter, it is possible to centrifuge our sample or to filter it (the latter suitable to the definition of what is in solution, e.g. < 0.45 μm, < 0.22 μm, etc.). Losses to, or gains from, any container's surface/vessel's walls or filter etc. are kept to a minimum by ensuring all apparatus is pre-cleaned and then rinsed or flushed with the liquid sample itself. Pre-concentration, if required for the later measurement technique, may be achieved using ion exchange-SPE cartridges to isolate the analytes contained in a large volume of original sample and then, after washing away any matrix on the SPE, collect the analytes in a suitable vessel by elution using a smaller volume of eluent to release them. The resulting solution sample may require acidification to pH ~2 (from an acid not containing any analyte of interest, e.g. do not use nitric acid!, or possible interference to the later measurement technique) to further stabilize the sample. It is noted that known analytical sample volumes before and after processing are identified and recorded throughout.

2. *A liquid, water-based sample requiring an organic chemistry measurement, that is quantitative, of molecular compounds; that is, organic nitrogen-containing compounds to be measured in solution from the tank.*

Being a water sample from a fish tank, we would need to consider processing and stabilizing our sample as soon as possible. The possibility of losses because of the presence of particulate matter absorbing, or light-activated degradation or biological activity using up, our nitrogen-containing analytes in the sample can be high; the integrity of the sample and the target analytes of interest would be lost. Freezing of the sample as soon as possible after collection while being kept in the dark may be considered if immediate processing is not readily available and would allow a delay. To isolate those analytes in solution from particulate matter, it is possible to centrifuge our sample or to filter it (the latter suitable to the definition of what is in solution, e.g. < 0.45 μm, < 0.22 μm, etc.). Losses to, or gains from, any container's surface/vessel's walls or filter etc. are kept to a minimum by ensuring all apparatus is made of the right non-absorbing materials for organic species, is pre-cleaned and then rinsed or flushed with the liquid sample itself. Pre-concentration from aqueous samples, if required for the later measurement technique, may be achieved using (i) a liquid–liquid extraction technique with a suitable non-polar organic solvent (immiscible with water) for dissolving the organic nitrogen-containing species or (ii) relatively non-polar/reversed phase-SPE cartridges to isolate the analytes contained in a large volume of original sample and then, after washing away the matrix, collecting the analyte(s)in a suitable vessel by elution using a smaller volume of organic solvent as eluent to release them. The resulting solution sample may require further stabilization by storage in the dark and at low temperature. It is noted that known analytical sample volumes before and after processing are identified and recorded throughout.

3. *A solid sample requiring an inorganic chemistry measurement, from a mixed inorganic and organic material that is quantitative, of an element; that is, the total nitrogen content to be measured in the sediment (mainly highly insoluble silicate and aluminosilicate mineral granules) from the tank which is contaminated with various organic components (fish food, fish faeces, dead organic matter, etc.).*

Only the total nitrogen content of the solid material present in the tank is required. The concentrations of the individual nitrogen-containing species are not asked for. At least two digestion approaches for a solid sample can be considered and the solid material suspended in the aqueous samples may also need to be included. This latter material can be collected by centrifugation and by filtration, as required. The settled solid, which is incorporated within the tank sediment, after suitable representative sampling, can be either considered as a wet sample or processed as the dry sample. Drying would need to ensure no loss of analyte occurred. Given the possible volatile nature of some analyte species under certain conditions of temperature and pH (e.g. organic nitrogen or NH_3/NH_4^+ at higher pH), some checks prior to a freeze-drying technique might be in order.

Known analytical quantities (weights) of material are taken and subjected to an acid digest both with and without oxidation. The former could include a hydrochloric acid and hydrogen peroxide digestion approach which would convert all the nitrogen containing species to NO_3^-. This means a simple total nitrate measurement later on will provide a total nitrogen value for the digest and hence the solid sample. The latter (digestion without oxidation and suitable reducing conditions) would allow conversion of all nitrogen species to NH_4^+, which may be targeted using a suitable analytical technique to provide a total nitrogen value from the digest.

As we will see in the next chapter, an instrumental separation technique based upon liquid chromatography can also be employed directly to the filtered and stabilised water samples 1 and 2 from above. This can separate nitrate, nitrite and ammonium ions and the organo-nitrogen compounds and then, by using a suitable detector, measure their concentration.

Instrumental measurement techniques

4

Learning outcomes

Once you have read through this chapter, worked through the activities and problems, and reviewed the feedback, you should be able to:

1. distinguish between **qualitative** and **quantitative analysis**;

2. understand the classification of different types of instrumental measurement techniques;

3. distinguish between the different requirements of an instrumental measurement technique for different types of samples and analyses;

4. state the details of a range of representative instrumental measurement techniques;

5. decide on an appropriate instrumental technique for making analytical measurements in a number of scenarios.

4.1 Introduction

This chapter deals with how we can identify the most appropriate measurement technique to use and the conditions to apply in order to generate the required basic data. However, it is important to understand at the outset that this is not intended to be a detailed description of all (or even most) of the instrumental techniques in existence. Instead, we will concentrate on some basic principles and use examples of a limited selection of instrumentation commonly encountered in most analytical laboratories to illustrate the main points.

4.2 **What is the measurement step?**

For the measurement step the analyst must attempt to identify the appropriate technique to be used in order to measure the property or analyte of interest in any given sample, and then perform this measurement. In order to do this you will have already evaluated the following three conditions:

1. Does the measurement need to be only **qualitative** (what is it) or also **quantitative** (how much is there)?

2. If quantitative, does it need to be at a **threshold** level (i.e. simply above or below a limit) or accurately and precisely known?

3. How **selective** must the measurement be in order to achieve the required level of **specificity** and freedom from interference?

Having identified these primary requirements for the measurement, which will normally include an estimate of the level of the analyte or property, the next stages in the process can be considered. This may include choices about the type of instrumentation that will meet the requirements, operating conditions and data acquisition procedures to be employed in a measurement step. A working calibration range of the technique should be identified, which will indicate if it is sufficiently sensitive, and with a low enough limit of detection for the analyte to be measured after any sample preparation. Now, that is a lot to consider and identify, which will be covered in Chapters 5–9, so for the moment we will confine ourselves to a consideration of various analytical techniques, some wet chemical but mainly instrumental, that are available to us, with some details that would help us make the right choices discussed above.

4.3 **Classification of analytical technique by underlying basis**

With the complexity of a discipline that has evolved over many years, the number of measurement techniques available in the analytical chemistry world has become extremely large. So, placing them into some sort of framework to facilitate selection for a particular task is of use. Table 4.1 identifies analytical techniques with respect to the underlying basis upon which the measurement is made.

Many of these techniques can be coupled together in order to increase the selectivity, sensitivity, detection limit, and working range. A good example of this is the coupling of a separation technique like **liquid chromatography** (LC) with a spectroscopic technique like mass spectrometry (MS), which revolutionized the use of MS for the determination of organic molecules in complex samples.

An overview of the techniques listed in Table 4.1 now follows. A detailed description of each is beyond the scope of this book; however, we will go on to describe several in more

Table 4.1 General classification of analytical techniques

Basis to technique	Technique	Notes
Chemical	Gravimetry Titrimetry	Classical methods based on the measurement of mass and volume
Biochemical	Enzymic Immunochemical	Biological methods based on the activity of enzymes for substrates and the affinity of antibodies
Electrochemical	Electrogravimetry Potentiometry Conductometry Voltammetry Coulometry	A range of methods based on the measurement of potential difference and current
Thermal	Thermogravimetry Differential thermal analysis Differential scanning calorimetry	Methods based on the measurement of changes in temperature, mass, and heat
Spectroscopic (electromagnetic)	Ultraviolet-visible Fluorescence Infrared Microwave Raman CHNS(O)	Molecular spectroscopy based on rotational, vibrational, and electronic energy transitions in molecules
	Atomic emission Atomic absorption Atomic fluorescence	Atomic spectroscopy based on electronic transitions of electrons in outer orbitals
	X-ray fluorescence	Transitions of inner orbital electrons
	X-ray diffraction	Structural, lattice formation
Spectrometric (mass)	Atomic	Mass spectrometry of elements
	Molecular	Mass spectrometry of organic compounds
Spectroscopic (magnetic)	Nuclear magnetic resonance	Interaction of certain nuclei with a magnetic field to create energy levels
	Electron spin resonance	Interaction of electrons with a magnetic field to create energy levels
Separation	Gas chromatography	Separation of volatile organic compounds by partition between a gaseous mobile phase and solid/liquid stationary phase
	Liquid chromatography	Separation of non-volatile organic compounds and inorganic ions by partitioning between a liquid mobile phase and solid stationary phase
	Electrophoretic	Separation of charged molecules and element ions by mobility in an electric field

detail to enable you to tackle the subsequent activities. These more detailed descriptions are included in Section 4.4; there are examples of **spectroscopic**[1] (atomic and molecular) and **separation** (chromatographic) techniques which are commonly encountered in most analytical laboratories.

4.3.1 Chemical

Gravimetric analysis involves the quantitative measurement of an analyte in a form that can be accurately measured by mass. It is a classical technique that requires the most important instrument in a laboratory (after the analyst themselves), which is an analytical balance. The analyte of interest is converted into a substance (of known composition) that can be separated from the sample and then weighed.

The sample containing the analyte, if solid, is usually transformed into a liquid state, often aqueous. A simple precipitation reaction is undertaken by adding a reagent to the sample solution, which forms an insoluble material of known composition with the analyte but not with other constituents in the sample. The precipitate, containing the analyte, is usually separated by filtration, washed free of any soluble impurities, dried or ignited to remove water, and then weighed. A well-known example of this technique is the determination of sulfate (SO_4^{2-}) in seawater by formation of the highly insoluble compound barium sulfate ($BaSO_4$) when a solution of $BaCl_2$ is added to the treated seawater sample.

Titrimetric analysis involves the quantitative measurement of an analyte in a reaction with a selective reagent that can be accurately measured by volume. It is a classical technique, called **volumetric analysis**, that utilizes a primary instrument known as a **burette**. A known quantity of sample containing the analyte, if in a solid state, is usually transformed into a liquid state, often aqueous. This is then titrated with a reagent of known concentration, added from the burette, to an end-point which is related to the analyte content. The end-point can be indicated visually (e.g. an acid/base or redox indicator) or by measurement (e.g. pH meter). Examples include the determination of the free fatty acid content of a foodstuff or a waste vegetable oil by neutralization using standard potassium hydroxide solution, and another would be the determination of extractable vitamin C from a food and its titration using standard dichlorophenol–indophenol (DCPIP) solution.

4.3.2 Biochemical

Enzyme-linked immunosorbent assay (ELISA) is a plate-based assay technique for liquid or wet samples designed for detecting and quantifying substances such as peptides, proteins, antibodies, and hormones. Other names, such as enzyme immunoassay[2] (EIA), are also used to describe the same technology. In an ELISA, an antigen[3] must be immobilized to a solid surface and then complexed with an antibody that is linked to an enzyme. Detection is accomplished by assessing the conjugated enzyme activity via incubation with a substrate to produce a measureable product, such as a colour change or fluorescence effect. The most crucial element

---

[1] The reader will encounter certain techniques presented in terms of their general study and use, and as such may be identified by the term 'spectroscopy'. However, they will also encounter techniques presented in terms of their use for identification and measurement of analytes, and as such can be identified by the term 'spectrometry'. Hence, the name given to many instruments is 'spectrometer'.

[2] Any type of assay that uses antibodies to measure the concentration of an analyte in a sample.

[3] Substance, protein, chemical compound, or virus that is able to elicit an immune response against which antibodies are raised.

of the detection strategy is a highly specific antibody–antigen interaction which is also highly sensitive. The most well-known example is the pregnancy test.

Other types of assays, similar to ELISA but not necessarily immunological in nature, can be performed using combinations of immobilized binding agents, at least one of which must be specific for the analyte (e.g. an enzyme or nucleic acid probe), and a detection agent to quantify. Examples include the determination of horsemeat in beef products and the veterinary drug phenylbutazone, which are banned for human consumption.

4.3.3 Electrochemical

Voltammetry is an electrochemical technique in which the current at an electrode is measured as a function of the potential, or voltage, applied to the electrode. The resulting current-potential display is called a voltammogram. There are several variations, including polarography (DC or differential pulse), anodic stripping, cathodic stripping, cyclic and linear sweep. Of these, anodic stripping voltammetry (ASV) and cathodic stripping voltammetry (CSV) are perhaps the most common, and are used to measure ions in solution. Typically a three-electrode system is used, comprising: a working electrode (often an Hg drop); a reference electrode; and a counter electrode. A voltage is applied to the working electrode (positive for anions and negative for cations) for a set period of time while it is immersed in the sample solution. Ions in the sample solution migrate towards the working electrode and become electroplated to it. The voltage is then reversed, positive to oxidize cations (ASV) or negative to reduce anions (CSV), which sequentially re-enter solution depending on their standard electrode potentials. At the same time, the counter electrode measures the current. The voltage at which they re-enter solution and the amount of current (peak size) provides information on which cation, and how much, is present in solution. Examples include the determination of potentially toxic metallic elements such as Pb in drinking waters or extracts from soil or sediment samples.

Potentiometry is a technique which measures the potential (voltage) of a solution between two electrodes. One electrode is usually a reference electrode held at a constant potential, while the other is a working electrode whose potential varies with the composition of the solution they are both immersed in. The potential difference between the electrodes provides a measure of composition. Modern analytical potentiometry usually employs a working electrode made selective for a particular ion of interest, known as an ion selective electrode (ISE). Examples include the common pH electrode and electrodes for the determination of fluoride in toothpaste or nitrate in river waters.

Electrogravimetry allows analyte ions in solution, usually metal or metalloid, to be deposited onto an electrode, usually a cathode, by the passage of current at a particular voltage. This allows selective reduction of the analyte ion to take place for a given time, similar to electroplating. The electrode is weighed before and after the process, in order to quantify the analyte. An example is the determination of copper or nickel ions in industrial waste solutions of metals through their deposition onto a pre-weighed platinum electrode.

Coulometry is the name given to a group of electrodic techniques in which the amount of electricity (the charge in coulombs) consumed or produced during an electrolysis reaction is measured. This is proportional to the concentration of analyte involved in the electrochemical reaction. Because the technique involves electron transfer, suitable analytes include those able to undergo redox reactions in solution.

There are two basic categories of coulometric techniques. In **potentiostatic coulometry** the electric potential is held constant during the reaction using a potentiostat to accurately quantify the concentration of a species. In **amperostatic coulometry** the current is kept constant using an amperostat to accurately quantify the concentration of a species. Samples are usually in the liquid state. Examples include the aqueous titrimetric measurements of electroactive species in solution such as the measurement of water content of butter, cheese, or even petroleum using the Karl Fischer reaction.

Conductometry is a measurement of electrolytic conductivity, often used to follow a chemical reaction. Conductometry is often employed to determine the total conductance of a solution, which is dependent upon its composition, or to analyse the end point of titrations that include ions. The end point is usually identified by a significant change in conductivity due to 'transport effects'. Samples are usually in the liquid or aqueous state. An example of the titrimetric conductivity technique would be a neutralization measurement between an acid and a base, where a change in the concentration of H_3O^+ ions at the end point results in a major change in conductivity.

4.3.4 Thermal

Thermogravimetric analysis (TGA) involves measuring the changes in physical and chemical properties of a sample (often loss in mass) in a controlled atmosphere as a function of temperature. The sample, usually a solid, is contained within a temperature controlled cabinet, on a special receptacle which is part of a high-precision mass-measuring device. As the temperature is raised, the change in mass of the sample is displayed as a thermogram. Any decomposition or change in the sample, in the presence of an inert or reactive atmosphere, can be followed in this way. The thermogram produced is specific to the material under investigation. The temperature range can be from below room temperature up to 2000°C. Kinetic changes, via thermograms at selected temperatures against time, can also be measured. Any evolved gases can also be analysed using infrared spectroscopy or mass spectrometry. For example, the technique can be used for investigation of the chemistry underlying the thermal stability and decomposition of compounds such as polymers or building materials.

Differential scanning calorimetry (DSC) is another thermo-analytical technique, but one that looks at how a material's heat capacity (Cp) is changed by temperature. In this, the difference in the heat required to increase the temperature of both a sample and a suitable reference material is measured as a function of temperature. The sample, usually a solid, and reference material both experience the same close temperature environment and the differences required to achieve this are plotted as the sample temperature increases, generally linearly as a function of time. Phase and state changes can be followed using this technique. Examples include the measurement of phase-properties in polymers and liquid crystals.

Differential thermal analysis (DTA) is used to measure the difference in temperature between a sample and a suitable reference material where both experience the same thermal environment. Modern day instruments incorporate this facility within DSC and TGA instruments.

4.3.5 Spectroscopic (electromagnetic) molecular

Ultraviolet (UV)-visible spectroscopy is based on the measurement of light, in the UV and visible part of the spectrum. Radiation is absorbed at a particular wavelength or wavelength range, by a single molecular compound or mixture, as a result of electronic transitions within molecular bonds. The amount of radiation absorbed is proportional to the concentration of the analyte, usually a polyatomic molecule or organometallic complex. Samples are usually in liquid form. Examples include the measurement of the copper concentration in coloured complex solutions and phosphate in river waters. This technique is discussed in more detail in Section 4.4.2.

Molecular fluorescence occurs when a molecule absorbs excitation radiation at a particular wavelength to reach an excited state. It then relaxes back to its lower state and re-emits radiation, but at a longer wavelength. The emitted radiation, measured at 90° to the excitation radiation to reduce noise, is proportional to the concentration of the analyte. The excitation process is an electronic transition within molecular bonds, usually those of organic molecules and organometallic complexes, so absorption of energy is usually from the UV region of the spectrum, with emission in the near UV and visible region of the spectrum. Samples are usually liquids. A good example of this, much used in laboratory practicals, is the measurement of quinine in tonic water. This technique is discussed in more detail in Section 4.4.2.

Infrared (IR) spectroscopy relies on the absorbance of IR radiation in the mid IR range of the electromagnetic spectrum, between ~2.5 and 25 µm. The wavelength is usually expressed in units of reciprocal cm, that is, $4000-400 \text{ cm}^{-1}$. In this range, the rotational-vibrational structure of covalent bonds in molecules can be examined. Different types of bond have a particular vibrational frequency so different molecules have unique absorbance (or transmittance) spectra. The technique is usually qualitative, but it can be used to quantify because absorbance is proportional to the quantity of material present. Samples may be measured in their solid, liquid or gaseous states. One example is the identification and determination of alcohol in blood.

Raman spectroscopy, like IR spectroscopy, is a molecular technique which measures the rotational-vibrational spectrum of a compound. Unlike IR, however, the sample is illuminated by an intense monochromatic light source, such as a laser, and the scattered light is measured as a function of wavelength and intensity. This is caused by polarizability of the electron density in the bond system rather than, as with simple IR spectroscopy, change in any permanent dipole moment present. Hence, non- or weakly absorbing bond systems in the IR range may be 'Raman' active and provide structural information. This qualitative technique can also be quantitative because the intensity of the scattered Raman radiation is proportional to the quantity of material present. A recent example of its use includes the remote detection of explosives using laser light.

Microwave spectroscopy is a technique in which information is obtained on the structure and chemical bonding of molecules and crystalline materials by measurement of the absorbance (or emittance) of selected wavelengths within the microwave range of the electromagnetic spectrum (~1 to 1000 mm). Instruments available usually work within the scanned frequency range 6–25 GHz. This range may also be used to study the fundamental rotational and rotational–vibrational structure of materials based upon the principle that all molecules can rotate in their gaseous and liquid states. This qualitative technique can also be quantitative because the absorbance (and emission) effect is proportional to the quantity of material present. One novel example includes the identification and measurement of ammonia in the interstellar medium.

4.3.6 Spectroscopic (electromagnetic) atomic

The atomic absorption spectrometric (AAS) techniques include flame AAS (FAAS) and electrothermal AAS (ETAAS). These are based on the absorption of radiation, at a specific wavelength, by analyte atoms either in a flame or in an electrically heated graphite tube. As the radiation passes through the flame or the graphite tube, some of the radiation is absorbed by the electronic ground-state analyte atoms. The change in intensity of the radiation passing through is proportional to the concentration of the analyte. Sample introduction is usually in liquid form. An example of its use is the measurement of Pb and Zn in sewage sludge and fluids. The technique of FAAS is discussed in more detail in Section 4.4.1.

The three general forms of **atomic emission** (AE) spectrometry encountered are flame (FAES) and microwave/inductively coupled plasma (MIP/ICP) techniques. The technique is based on the **emission** of radiation, at specific wavelengths, by analyte atoms in a high temperature environment. This can be produced by a flame or a plasma (a partially ionized gas). The electronic excitation of atoms is thermally initiated and it is the intensity of the emitted line radiation that is measured. This is proportional to the concentration of the analyte. Sample introduction can be in mainly liquid, but also gaseous phases. Examples include the determination of Li, Na, and K in body fluids using FAES; the measurement of halogens in organic compounds by MIP-AES and the determination of metals, metalloids and selected non-metals in extracts from waste materials by ICP-AES. The technique of ICP-AES is discussed in more detail in Section 4.4.1.

Atomic fluorescence is where light from a high intensity line source is absorbed by an atomic species present in a flame or vapour state. This absorbed light excites an electron in the atom to a higher energy level. The electron is not stable at this higher energy level and it returns to the ground state losing its excess energy in the form of light. The light emitted is at the same wavelength as that absorbed, but measurements are taken at 90° to prevent confusion between incident and emitted radiation. The intensity of this re-emitted light is proportional to the concentration of that atomic species. Sample analytes are determined in their gaseous state, but liquids may also be introduced. Examples include the determination of low levels of elements like arsenic, selenium, antimony, and mercury in foodstuffs.

X-ray fluorescence spectroscopy is a technique whereby X-rays illuminate a sample and, being of higher energy, they cause an electron from an inner orbital of an analyte atom to be ejected. An electron in an orbital further out fills this hole and radiation characteristic of the energy gap between the two orbitals is emitted, that is, energy specific to individual analytes, which is itself in the X-ray region of the spectrum. The amount of radiation emitted is proportional to the concentration of that element present. Two main variants of instrumentation exist: energy dispersive and wavelength dispersive. Hand-held (portable) energy dispersive instruments are also available. Samples are usually presented as solids or liquids. A novel example includes the forensic elemental study of paint pigments in 'old master' paintings.

Energy-dispersive X-ray spectroscopy (EDXS), also known as energy dispersive X-ray analysis (EDXA), is an elemental analytical technique usually associated with scanning electron microscopy (SEM). The high energy electron beam used in SEM can cause an electron in an inner orbital to become excited and ejected. An electron in an orbital further out fills this hole and radiation characteristic of the energy gap between the two orbitals is emitted. This radiation has energy, in the X-ray region of the spectrum, specific to the analyte, and its intensity is proportional to the concentration. Samples are usually presented in solid form to the SEM where the EDXS mode is selected. An example is the measurement of surface contamination on semiconductor wafers.

4.3.7 Spectroscopic (electromagnetic) structural

X-ray diffraction (XRD) analysis, also known as X-ray crystallography, is a technique used for determining the atomic and molecular structure of a solid crystalline material. A beam of monochromatic or narrow bandwidth X-rays are directed at the material sample where the atoms within the solid lattice cause them to be diffracted in specific directions, which are read as angles relative to the incident beam. The pattern of angles, shown as 2θ (2 theta), is unique to a particular structural arrangement and is used as a qualitative measurement. The intensity of the primary diffracted beams at their particular angle may be considered proportional to the concentration of the material present in crystalline mixtures under suitable conditions. An example of its use would be for the identification and estimation of the components in a calcite ($CaCO_3$) or dolomite [$(CaMgCO_3)_2$] ore mixture.

4.3.8 Spectrometric (mass)

Atomic mass spectrometry is a technique based upon the measurement of atomic ions. The ions are usually formed in a high temperature environment. It has several variants, including inductively coupled plasma mass spectrometry (ICP-MS), thermal ionization mass spectrometry (TIMS), accelerator mass spectrometry (AMS), and secondary ion mass spectrometry (SIMS). The technique of ICP-MS is commonly found in analytical labs. This produces singly-charged, elemental analyte ions with mass-to-charge ratio (m/z) equivalent to their isotopic masses, for example, copper will give a signal at m/z 63 and 65. There are several types of ICP-MS, including quadrupole (unit resolution), sector field (to allow higher resolution), and time-of-flight (higher mass filtering capabilities) mass spectrometry. Each one separates the ions according to their m/z and counts the number of each ion reaching the detector. The number of counts is proportional to the concentration of each mass. The sample is usually introduced as a liquid or solid (e.g. after laser ablation), but gases may also be analysed directly. The study of geological materials and their elemental make-up is a typical example.

Molecular mass spectrometry relies on softer ionizing sources, so that molecules are fragmented incompletely, or not at all, and ionized to varying degrees. The fragmentation pattern enables structural information to be elucidated and the identification of unknown species in samples. Types of ionization available include electron ionization, chemical ionization, fast atom bombardment, electrospray ionization, and matrix-assisted laser desorption/ionization (MALDI). Mass spectrometers are similar to those described for atomic mass spectrometry, with the possible addition of several quadrupoles or ion traps which may also be used for secondary fragmentation and mass selection. Molecular mass spectrometry is often used in combination with gas or liquid chromatography separation techniques so that on-line quantification and identification may be achieved. Samples can be introduced in gaseous, liquid, or solid form and a classical example might be the determination of drugs in an athlete's body-fluid sample.

4.3.9 Spectroscopic (magnetic)

Nuclear magnetic resonance (NMR) spectroscopy is the definitive analytical technique for determining molecular structure. It makes use of the fact that certain atomic nuclei have a ground state nuclear spin, I, of ½, the most commonly observed being hydrogen (^1H) and carbon (^{13}C) for the structural analysis of organic compounds. However, the NMR signal from over 20 other nuclei can also be readily observed with the appropriate hardware. The NMR signal is detected in the presence of a strong external magnetic field, so most modern NMR spectrometers utilize specialized superconducting magnets cooled with liquid helium. Samples for NMR are usually dissolved or diluted in deuterated solvents contained in a narrow glass tube, though solids and gases can also be analysed using specialized hardware. The sample is placed in a constant, homogeneous high-strength magnetic field which induces individual nuclei within the atoms of the molecule to align with or against the direction of the field. A pulse of radiofrequency (RF) energy is applied, thereby effecting a change in the orientation of NMR-sensitive isotopes within the magnetic field. Following this 'resonance' state the energized isotopes 'relax' back to their original orientation. It is the small energy difference between the resonance state and the relaxed state that is observed to give an NMR spectrum. This spectrum, which demonstrates peaks, is calibrated with an internal standard, either tetramethylsilane (defined chemcial shift of 0), or more usually the residual proton signal of the deuterated solvent is used and the positions of the peaks are expressed in terms of a **chemical shift**. The power of NMR spectroscopy is that the resonance frequency of an individual nucleus is dependent on its molecular environment so is unique to a particular material. The spectrum shows peaks, whose number, position, and intensity allows an unambiguous confirmation of molecular structure. An example includes where a synthetic organic compound for pharmaceutical use has been produced and both confirmation of form and purity is required.

Electron spin resonance (ESR) or electron paramagnetic resonance (EPR) spectroscopy is used to study paramagnetic materials (those with unpaired electrons). The basic concepts of EPR are analogous to those of NMR, but it is the spins of electrons that are excited instead of atomic nuclei. Microwave energy is used to induce electron energy changes within an external strong magnetic field and the differences in electronic environment that exist for the unpaired electrons are measured. The position, pattern, and intensity of peaks in the spectrum provide both spectroscopic and structural information about the sample. The EPR technique is particularly useful for studying metal complexes or organic radicals and is used in dating certain materials. Both solid and liquid samples can be measured.

4.3.10 Elemental

A **CHNS elemental analyser** is used to determine the elemental content of carbon, hydrogen, nitrogen, and sulfur in a sample. Some instruments allow a measurement of oxygen (CHNS/O) as well. From this the mass percent of C, H, N, S, and O can be determined for a variety of sample types. This includes in many cases organic compounds, thus providing additional information such as purity, empirical formulae, etc. The instrument uses a combustion process to oxidize the sample into simple compounds (CO_2, CO, N_2, H_2O, etc.) which can then be separated and quantified. Sample sizes can be as small as the lower mg range and are generally introduced as solids or non-volatile liquids. The organic content of a soil or sediment sample can be determined by this technique.

4.3.11 Separation

Chromatographic techniques are a means of separating a mixture of compounds or ions (the analytes) into individual components. Modern instrumental techniques utilize a column which has a **stationary phase** immobilized within it and a **mobile phase** which passes through it. Analytes in the sample mixture interact with the stationary and mobile phases to different degrees, depending on their properties, as they pass through the column. This partitioning between phases causes the analytes to become separated, as they are carried by the mobile phase and eluted from the end of the column at different times. The individual analytes are then measured using a suitable **detector** at the end of the column. The two major chromatographic techniques encountered in the laboratory are gas chromatography and high performance liquid chromatography, which are described in more detail in Section 4.4.3.

In **gas chromatography** (GC), the gaseous or liquid sample, usually organic, is injected into the head of the chromatographic column, which contains the stationary phase, and is vaporized. The sample is transported through the column by the flow of an inert, gaseous mobile phase and the components separated on the basis of their boiling point and polarity. The separated species require a suitable detector for measurement. This technique, used with mass spectrometry as a detector, can be used in the forensic analysis of blood samples for alcohol determination in court cases. This technique is discussed in more detail in Section 4.4.3.

High performance liquid chromatography (HPLC) is the other major separation technique in common use. It can be used to separate and determine a wide range of analytes in a variety of liquid samples. It is complementary to GC because it can be used for analytes that are non-volatile and/or polar. A solution of the sample to be analysed is injected into a liquid mobile phase, which is pumped (under pressure to achieve a reasonable flow rate) through the column. This column contains particulate matter with specific chemical characteristics which interact with the individual analytes to different degrees. This allows their separation and detection for measurement. The determination of anions such as Cl^-, Br^-, F^-, SO_4^{2-}, PO_4^{3-}, NO_3^-, etc. in drinking water or vitamins in food extracts are just two examples of the technique's use. This technique is discussed in more detail in Section 4.4.3.

Electrophoresis is used to separate mixtures of charged molecules or elemental ions according to their mobility within an electric field and the ability to partition into particular phases or zones. The two major forms of this technique encountered are **gel** and **capillary zone electrophoresis**. Samples are generally in a liquid state. In gel electrophoresis, a medium density electric field is used to separate large charged molecules (frequently biological) in a conducting polymeric gel. In capillary zone electrophoresis a high density (voltage) electric field is used to separate ionic species in a capillary column containing a conducting liquid. The separated species require a suitable detector for measurement. Gel electrophoresis is employed in the analysis of nucleic acids (DNA, RNA) and proteins. Capillary electrophoresis can also be used for this and also for the separation of simple ions.

4.4 Detailed descriptions of selected techniques

4.4.1 Atomic spectrometry

Table 4.2 lists the type of radiation associated with particular named regions of the electromagnetic spectrum together with the range of wavelengths and frequencies broadly identified with those regions, and the type of energetic transition that can occur when a photon in these spectral ranges interacts with matter.

The interactions shown in Table 4.2 make it possible to detect and quantify elements and molecules by utilizing particular regions of the electromagnetic spectrum. Atomic spectrometry makes use of the UV and visible part of the spectrum. Two element-specific techniques of atomic spectrometry are described in the following sections.

4.4.1.1 Flame atomic absorption spectrometry (FAAS)

This technique is based on the **absorption** of radiation, at a specific wavelength, by analyte atoms in a flame. A schematic diagram showing the basic elements of a flame atomic absorption spectrometer is shown in Figure 4.1. The light source is a hollow cathode lamp (HCL) which emits radiation from a cathode made from the analyte of interest, usually a metal or metalloid but some non-metals can also be measured. A large range of such lamps are available that include the vast majority of the elements of general analytical interest. The radiation from the HCL will be most intense at the same wavelengths that the analyte in the flame will selectively absorb. As the radiation passes through the flame, some of the radiation is absorbed by the ground-state analyte atoms, and then carries on through the optical system of the spectrometer.

Table 4.2 The electromagnetic spectrum and interactions of radiation with matter

Type of radiation	Wavelength range/m	Frequency range/Hz	Interaction
Gamma rays	$< 10^{-11}$	$>10^{19}$	Nucleus of an atom
X-rays	$(10 \text{ to } 0.01) \times 10^{-9}$	$30 \times 10^{15} \text{ to } 3 \times 10^{19}$	Inner shell electrons of atom
Ultraviolet	$(380 \text{ to } 10) \times 10^{-9}$	$790 \times 10^{12} \text{ to } 30 \times 10^{15}$	Outer electrons of atom
Visible	$(760 \text{ to } 380) \times 10^{-9}$	$(400 \text{ to } 790) \times 10^{12}$	Outer electrons of atom
Infrared	$(1000 \text{ to } 0.76) \times 10^{-6}$	$(0.3 \text{ to } 400) \times 10^{12}$	Outer electrons of some atoms of lower energy; Vibrational and rotational-vibrational states of molecules
Microwave	1 to 0.001	$(0.3 \text{ to } 300) \times 10^{9}$	Rotational states of molecules; Spin states* of the electron
Radio wave	10^{3} to 1	$3 \times 10^{5} \text{ to } 300 \times 10^{9}$	Spin states* of the nucleus

*For energy levels split by a magnetic field.

The sample, dissolved in a suitable solvent, is formed into an aerosol (i.e. nebulized) and fed into the gas stream at the base of the burner and thence into the flame (air/acetylene, nitrous oxide/acetylene or air/propane). The light from the HCL, having passed through the flame and been partly absorbed there, can be focussed directly onto a diffraction grating by means of a spherical mirror. The diffraction grating can be set to isolate a particular wavelength that

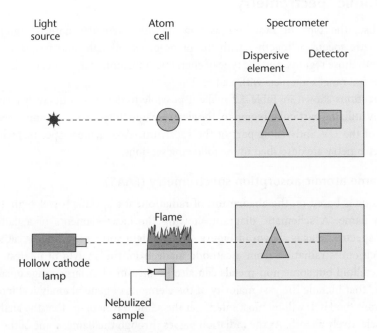

Figure 4.1 Schematic of a flame atomic absorption spectrometer

is characteristic of the element being measured. After leaving the grating, light of the selected wavelength is focussed onto the photocell and detected.

Because the HCL is element-specific and produces radiation of exactly the right energy for ground-state analyte atoms to absorb, AAS is sometimes referred to as a 'lock and key' system. For example, if you want to measure Ca in the flame you use a Ca HCL, and for Mg you would use an Mg HCL, and so on. The radiation emitted from the HCL is measured after passing through the flame in the presence of calibration solutions (known concentrations of the analyte) and samples. The more analyte atoms in the flame, the greater the absorbance of radiation and the intensity goes down. Absorbance is expressed as the ratio of the intensities of the incident and transmitted radiation and calibrated against concentration:

$$A = \log_{10}\left(\frac{I_i}{I_t}\right) \qquad (4.1)$$

where A = absorbance, I_i = intensity of incident radiation, and I_t = intensity of transmitted radiation.

The linear working range is dependent upon the element but is often around two orders of magnitude concentration.

4.4.1.2 Inductively coupled plasma atomic emission spectroscopy (ICP-AES)

This technique is based on the **emission** of radiation, at specific wavelength, by analyte atoms in a high temperature plasma (a partially ionized gas). Like FAAS, the electronic excitation of atoms is necessary, but this time rather than measure the absorbed radiation it is the emitted radiation that is measured, and the inductively coupled plasma (ICP) serves as both atom cell and the light source. This means that, unlike for FAAS, there is no need to use an HCL as a source of radiation. However, in order to provide sufficient excitation of atoms, it is necessary to have a much hotter environment than can be provided by a flame. This problem is addressed by using an ICP (Figure 4.2), which can be as hot as 7000 K. In this hot plasma, atoms and ions are easily excited to higher energy electronic states. When they relax to the ground state they emit radiation at a wavelength characteristic of that particular element. The intensity of this emission is therefore proportional to the concentration of the element in the plasma and hence to any sample directly introduced to it:

$$I_a = I_s - I_b \qquad (4.2)$$

$$I_a = Kc \qquad (4.3)$$

where I_a = corrected analyte emission, I_s = intensity of emission at analyte wavelength, I_b = intensity of background emission at analyte wavelength, c = analyte concentration, and K = proportionality constant related to sensitivity.

Samples are introduced into the ICP as liquid **aerosols** produced by a **nebulizer**. The sample aerosol droplets pass through a quartz chamber called a **spray chamber** and thence into a **quartz torch** where the plasma is produced and maintained. The torch is formed from three concentric tubes through which flows argon gas; the central tube carrying the sample aerosol. The torch is surrounded by a three-turn, water-cooled copper coil through which an RF current is passed. This alternating RF current sets up a magnetic field around the coil so that when the argon gas is first ionized the electrons and ions produced circulate in the field and start up a chain reaction, causing a high-temperature plasma to form which is sustained as long as the RF field is present. The high temperature causes the sample aerosol droplets to vaporize,

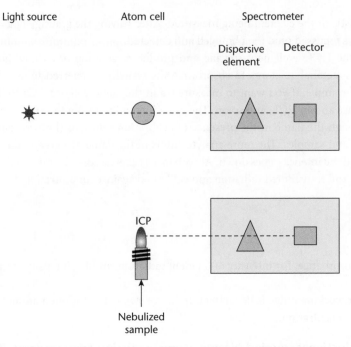

Figure 4.2 Schematic of an inductively coupled plasma atomic emission spectroscopy instrument

decompose, atomize, and ionize. The atoms and ions become excited (electronic transitions to higher energy levels) and these excited species then return to their lower energy states by giving off radiation at the characteristic wavelengths of the elements involved. The radiation passes through light gathering optics into a **spectrometer** which is able to separate the wavelengths of emitted radiation using a dispersive element called a **diffraction grating**. The separated wavelengths then fall upon a detector, usually an array of semiconductor **photodetectors** such as charge coupled devices (CCDs) or charge inductive devices (CIDs). In this way the intensity of the radiation at all of the separate wavelengths can be measured simultaneously, between ~167 and 766 nm, allowing the instrument to analyse for every element at once. The intensity of each line is then compared with previously measured intensities of known concentrations of the elements (calibrants), and their concentrations are computed (see Section 5.4). This sensitive, multi-element, element-specific instrumentation is very efficient, covering some five orders of linearity and can measure down to low $\mu g\ L^{-1}$ levels.

4.4.2 Molecular spectroscopy

Molecular spectroscopy encompasses a range of instrumental methods predicated on the interaction of electromagnetic radiation (Tables 4.2 and 4.3), but with molecules rather than atoms. Sections of molecules which can undergo detectable electron transitions in the 'UV to visible' part of the spectrum can be referred to as **chromophores**. By way of example we will describe only UV-visible and molecular fluorescence spectrometry, two techniques which are available in most analytical laboratories.

4.4.2.1 **UV-visible spectroscopy**

UV-visible (or UV-vis) spectroscopy is based on the measurement of light (UV and visible wavelengths) absorbed at a particular wavelength by a particular molecular compound or range of compounds. It involves electronic transitions within molecular bonds, usually those of organic molecules and metallic complexes.

Table 4.3 Electronic transitions associated with different organic molecular bonds

Chromophore	Example	Excitation	λ_{max}, nm	ε	Solvent
C=C	Ethene	$\pi \rightarrow \pi^*$	171	15,000	hexane
C≡C	1-Hexyne	$\pi \rightarrow \pi^*$	180	10,000	hexane
C=O	Ethanal	$n \rightarrow \pi^*$	290	15	hexane
		$\pi \rightarrow \pi^*$	180	10,000	hexane
N=O	Nitromethane	$n \rightarrow \pi^*$	275	17	ethanol
		$\pi \rightarrow \pi^*$	200	5,000	ethanol
C–X					
X = Br	Methyl bromide	$n \rightarrow \sigma^*$	205	200	hexane
X = I	Methyl iodide	$n \rightarrow \sigma^*$	255	360	hexane
Phenyl-	Aniline	$\pi \rightarrow \pi^*$	230	8000	hexane
			280	3000	

UV-vis spectroscopy is more sensitive than titrimetry, being capable of measuring the quantities of analytes down to the μg L^{-1} range. Many spectrophotometric methods utilize the ability of transition metals to form coloured (and therefore light absorbing) compounds with organic compounds, in order to measure the transition metal concentration. An example is shown in Figure 4.3 of the cuprizone complex of Cu, where λ_{max} (the wavelength of maximum absorbance) is at 600 nm. This is the wavelength chosen to undertake measurements of absorbance because it will give maximum sensitivity.

The theory of quantitative determination by UV-vis spectroscopy is based on the Beer–Lambert law, which states that absorbance is proportional to concentration:

$$A = \varepsilon c l \tag{4.4}$$

where ε = the molar absorptivity constant, c = the concentration of the absorbing species (in mol L^{-1}), and l = the path length, that is, the distance the light travels through the solution (in cm). Comparison of the absorbance of a solution containing an unknown concentration of a compound with solutions containing known concentrations can be used to compute the concentration in the unknown. When the molar absorptivity constant ε is a large value, the sensitivity of the absorbance technique is also high. A schematic of the main elements in an instrument is shown in Figure 4.4.

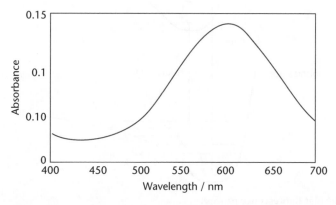

Figure 4.3 Absorbance of the copper(II)–biscyclohexanone oxaldihydrazone (cuprizone) complex in alkaline solution

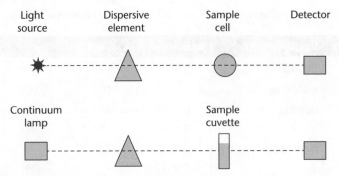

Figure 4.4 Schematic of a UV-visible spectrophotometer

It is possible to use UV-vis spectroscopy partly selectively, by choosing absorption wavelengths particular to a target bond type, such as aromatic compounds or even selected ions like nitrate. Unfortunately, the selectivity is limited to a range of compounds with similar types of bond unless the analyst is aware of all the other compounds present in the sample. However, when combined with a separation technique (e.g. HPLC and SPE) the selectivity is greatly increased because analytes with the same type of bond are separated before detection.

4.4.2.2 Molecular fluorescence spectrometry

Molecular fluorescence occurs when a molecule absorbs radiation to reach an excited state, then re-emits, at a different wavelength, when it relaxes back to its previous state. The excitation process is an electronic transition so absorption of energy is usually from the UV region of the spectrum, as shown in Figure 4.5.

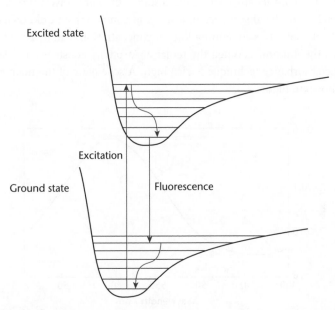

Figure 4.5 Molecular fluorescence process

After excitation, the molecule quickly relaxes to the lowest vibrational level of the excited electronic state, losing energy non-radiatively by transfer to solvent molecules. Fluorescence occurs when the molecule relaxes from this lowest vibrational level of the excited electronic state to the ground electronic state by the emission of radiation. The molecule can return to the ground state without fluorescence occurring, through collisions with molecules of solvent or other fluorescing species. The more flexible the molecule is the easier it is for collisional deactivation to occur, thus in most cases only rigid molecules will fluoresce.

In general, substances that absorb strongly in the UV, usually at wavelengths longer than 230 nm, may fluoresce, which is the case for many organic compounds. Those most likely to fluoresce have cyclic structures containing conjugated double bonds. As a result, most polynuclear aromatic hydrocarbons fluoresce strongly. The presence of certain groups such as $-OH$, $-OR$, or $-NH_2$ can increase fluorescence intensity, where R is an organic substituent, such as a methyl, ethyl, or phenyl group, etc. However, the presence of certain groups such as $-COOH$, $-NO_2$, or $-SO_3H$ can reduce or even eliminate fluorescence. The effect from closing a ring in a compound, for example as seen in the formation of a metal chelate, enhances fluorescence.

Different chemical species have characteristic excitation and fluorescence spectra. In order to measure fluorescence, it is necessary to have a source that emits UV radiation, a detector that responds to UV and visible radiation, and a means for separating the excitation and fluorescence radiation. A simple schematic of this arrangement is shown in Figure 4.6. It is worth noting that, unlike in UV-vis spectroscopy, the emitted radiation (transmitted radiation in UV-vis) is observed at a right angle to the beam of exciting radiation from the continuum lamp. This is to reduce the amount of scattered radiation detected, hence reducing the background noise and lowering the limit of detection.

It is noted that we use this technique in Example 4.3, coupled with HPLC, to identify the different polycyclic aromatic hydrocarbons (PAHs) and measure their concentrations.

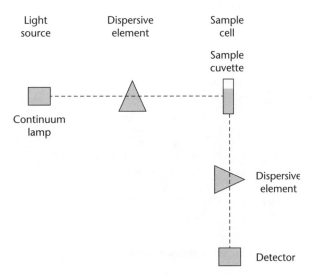

Figure 4.6 Schematic of a molecular fluorescence instrument

4.4.3 Chromatography

Chromatography is a **separation technique** where a number of analytes distribute between two phases, the mobile and stationary phases, over time. The stationary and mobile phases are chosen so that they have contrasting physical and chemical properties, so the analytes will distribute between them depending on their affinity for one or other of the two phases. The two major chromatographic techniques generally encountered are GC and LC.

4.4.3.1 Gas chromatography

In GC, the sample, usually organic in origin, is injected onto the head of the chromatographic column, which contains the stationary phase, and is vaporized. The sample is transported through the column by the flow of an inert, gaseous mobile phase (Figure 4.7).

The carrier gas is usually nitrogen, helium, hydrogen, argon, carbon dioxide, or even air, depending on the type of detector used. Small volumes of liquid samples are injected into a sample port at the head of the column where all of the components are immediately vaporized. The temperature of the sample port is usually about 50°C higher than the boiling point of the least volatile component of the sample. In split-less injection, all of this vapour is swept into the column by the carrier gas. In a split injection only part of the sample enters the column, the rest being vented; this is to avoid diluting concentrated samples or overloading a column. Most analytical columns in common use are **capillary columns** made from silica. These have an internal diameter usually between 100 and 300 µm, and can be many metres long. The stationary phase is coated onto the inside of the column as a thin layer. There are also **packed columns**, which have an internal diameter of between 2 and 4 mm, and are packed with finely divided, inert, solid support material (commonly based on silicaceous materials) coated with liquid stationary phase.

In order to separate the analytes reproducibly the column temperature must be controlled to within tenths of a degree. This can be at a constant or gradually increasing (stepped or gradient) temperature. A typical chromatographic run can take between 2 and 30 min. Stepped or gradient temperature programming is used to improve the speed and resolution of the separation when the sample has a wide boiling range. As the separated analytes exit the column they are detected one after another to give a chromatogram of detector response versus time (Figure 4.8).

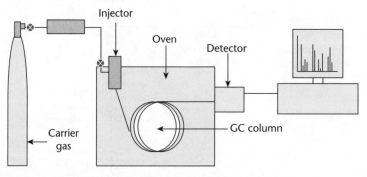

Figure 4.7 Schematic of a gas chromatography instrument

Peaks

1. 2,4,5,6-Tetrachloro-*m*-xylene (SS)
2. α-BHC
3. γ-BHC
4. β-BHC
5. δ-BHC
6. Heptachlor
7. Aldrin
8. Heptachlor epoxide (isomer B)
9. *trans*-Chlordane*
10. *cis*-Chlordane*
11. Endosulfan I

12. 4,4'-DDE
13. Dieldrin
14. Endrin
15. 4,4'-DDD
16. Endosulfan II
17. 4,4'-DDT
18. Endrin aldehyde
19. Endosulfan sulfate
20. Methoxychlor
21. Endrin ketone
22. Decachlorobiphenyl (SS)

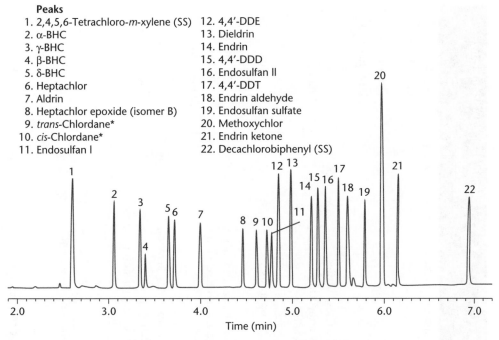

Figure 4.8 GC chromatogram of a mixture of pesticides (Source: Reproduced by permission of Restek Corporation.)

The most common detector is the flame ionization detector (FID), which responds to all organic compounds so is **non-selective**. Other detectors are more **selective** and respond to a range of compounds with a common physical or chemical property; for example, the electron capture detector is selective for certain functional groups. A detector such as a mass spectrometer is both **universal** (can detect all organic compounds) and selective (can identify ions of a specific *m/z*). Detectors can also be grouped into concentration dependent detectors and mass flow dependent detectors. The signal from a concentration dependent detector is related to the concentration of solute in the detector, and does not usually destroy the sample. Dilution with make-up gas (a gas flow that is used to sweep components through a detector to minimize band broadening) can lower the detector response unless carefully chosen. Mass flow dependent detectors usually destroy the sample, and the signal is related to the rate at which solute molecules enter the detector. The response of a mass flow dependent detector is unaffected by make-up gas. Table 4.4 presents a range of detectors used in GC systems together with their selectivity, response, and working range.

4.4.3.2 High performance liquid chromatography

HPLC is the other major separation technique in common use. It can be used to separate and determine a wide range of analytes in a variety of samples. It is complementary to GC because it can be used for analytes that are non-volatile and/or polar. A solution of the sample to be analysed is injected into a liquid mobile phase, which is pumped through the column (under pressure to achieve a reasonable flow rate) and thence to a detector. The stationary phase is held in a stainless steel or polymer column. The analytes partition between the mobile phase and the stationary phase to varying extents depending on their physical and chemical properties. If an analyte is strongly retained by the stationary phase it moves slowly with the flow of the mobile phase and *vice versa*. In this way, analytes separate into discrete bands that can be detected qualitatively or quantitatively with an appropriate detector, as shown in Figure 4.9.

Table 4.4 Gas chromatography detectors, their selectivity, and working range

Detector	Detector response	Support gases	Selectivity	Limit of detection	Working range
Flame ionization (FID)	Mass	Hydrogen and air	Most organic compounds	100 pg	10^7 (linear to 10^7)
Thermal conductivity (TCD)	Concentration	Reference	Universal	1 ng	10^5 (linear to $<10^5$)
Electron capture (ECD)	Concentration	Make-up	Halides, nitrates, nitriles, peroxides, anhydrides, organometallics	50 fg	10^5 (linear to 10^4)
Nitrogen–phosphorus	Mass	Hydrogen and air	Nitrogen, phosphorus	10 pg	10^5 (linear to 10^5)
Flame photometric (FPD)	Mass	Hydrogen and air ($\sim$oxygen)	Sulfur, phosphorus, (tin, boron, arsenic, germanium, selenium, chromium)	100 pg	$S = 10^3$ $P = 10^4$ (P linear to 10^4)
Photo-ionization (PID)	Concentration/mass	Make-up	Aliphatics, aromatics, ketones, esters, aldehydes, amines, heterocyclics, organosulfurs, some organometallics	2 pg	10^7 (linear to 10^6)
Electrolytic conductivity	Mass	Hydrogen, oxygen	Halide, nitrogen, nitrosamine, sulfur	1 pg	10^6 (linear to 10^5)
Mass spectrometer	Concentration/mass	n/a	All organic compounds	0.1 pg	10^6 (linear to 10^6)

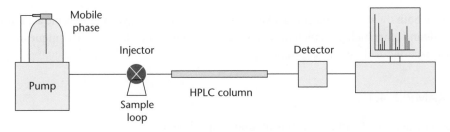

Figure 4.9 Schematic of a high performance liquid chromatography instrument

The stationary phase is commonly comprised of particles, between 3 and 10 µm in diameter, packed into a column of between 5 and 25 cm long and ~4–5 mm internal diameter. The particles are coated with an appropriate stationary phase so that the combined surface area for interaction with the analytes is extremely high as they travel through the column. The different types of HPLC available are given in Table 4.5.

One brief example for consideration is the determination of vitamin C in foodstuffs. HPLC has a number of advantages over older titrimetric methods including selectivity, reproducibility, and short analysis times. In addition, it is possible to devise methods that simultaneously

Table 4.5 Types of HPLC generally encountered together with analyte type and detection systems employed

Technique	Selectivity basis	Detector selectivity
Reversed phase (RP)	Lower polarity molecular species Mobile phase is more polar than stationary phase	Refractive index < conductivity < UV-vis and IR absorbance < molecular fluorescence < mass spectrometry
Normal phase (NP)	High polarity molecular species Mobile phase is less polar than stationary phase	As above
Chiral phase	Chirality of molecules	As above
Gel permeation (size exclusion)	Molecular size	As above
Normal phase ion chromatography	Separation of high polarity and ionic molecular species and element ions, e.g. F^-, Cl^-, Br^-, I^- and simple metal/metalloid ions in solution.	Refractive index < conductivity < UV-vis and IR absorbance < (molecular fluorescence and direct mass spectrometry—very limited usage) Can be linked to plasma techniques (OES and MS) when certain elemental constituents present, e.g. organometallic/ organometalloids, for very high selectivity speciation measurements

estimate both the basic L-ascorbic acid and its easily oxidized dehydro-L-ascorbic acid moiety in foods and other biological materials making it superior to the titrimetric technique in which only L-ascorbic acid can be determined.

EXAMPLE 4.1

Determination of anthocyanidins in food supplements
We can use the HPLC separation technique with a UV-vis detector to identify and measure the natural products in food supplements. As an example, we will concentrate on aqueous extracts containing anthocyanidins and their related compounds (anthocyanins). The most common anthocyanidins are cyanidin (red-purple), delphinidin (blue-purple), malvidin (deep purple), peonidin (red), petunidin (purple), and pelargonidin (orange-red), their structures being based upon the compound shown in Figure 4.10.

We can extract these compounds from the supplements using an aqueous extraction procedure similar to that used to extract the colourants from candies, discussed in Chapter 1. The extract from step 3 of the sample preparation process, contains various compounds that need to be separated prior to detection. In this case, we have decided to use an HPLC separating column that contains a stationary phase (a C_{18} type) with a strong affinity for non-polar compounds and a more polar mobile phase that passes through it. Compounds will partition themselves according to their individual affinity for the stationary phase and will separate out as they go through, being carried by the mobile phase, until arriving at the detector at different times. The detector we will use is a UV-vis spectrometer. We inject 10–20 μL of the aqueous extract onto our column and use a mobile phase of acidified methanol and acetonitrile to elute the compounds one after the other so that we end up with a separation, over time, of the various components of the mixture. The time taken to arrive at the detector, referred to as the retention time R_t, is dependent upon the compound itself and the response (peak area) is dependent upon how much of the compound is present. The effect is similar to that shown in Figure 4.11 using known concentration standards where the R_t values from the standards and their individual peak response would be used to calibrate the measurement system.

The responses from the samples would be compared with the standards for preliminary identification from any coincident R_t values and their peak areas (via their individual calibrations) to calculate their concentration in the extracts.

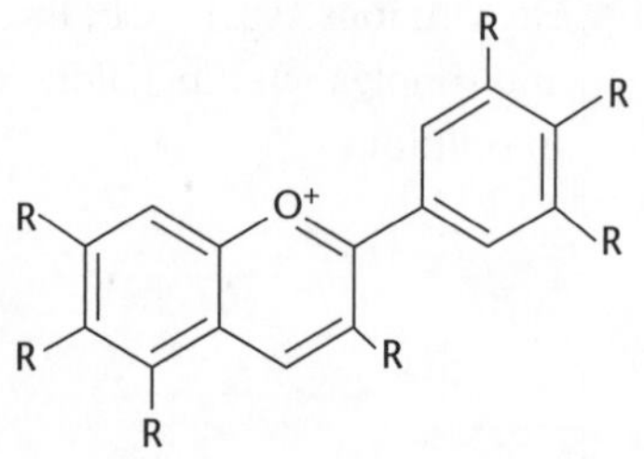

Figure 4.10 Base structure of anthocyanidin compounds

Column:	Acclaim RSLC 120, C18, 2.2 μm, Analytical (2.1 x 150 mm)	Peaks:		Conc. (μg/mL)
Eluent:	A: 10% Formic acid		1. Dp3Gal	—
	B: 10% Formic acid, 22.5% methanol, 22.5% CH_3CN		2. Dp3Glu	20.1
			3. Cy3Gal	11.0
Gradient:	0.0–12.0 min, 9% B		4. Dp3Ara	—
	12.0–25.0 min, 35% B		5. Cy3Glu	12.3
	25.0–30.0 min, 50% B		6. Pet3Gal	—
	30.0–35.0 min, 9% B		7. Cy3Ara	—
Flow Rate:	0.475 mL/min		8. Delphinidin	—
Inj. Volume:	2.0 μL		9. Pet3Glu	9.91
Temp.:	35 °C		10. Peo3Gal	3.91
Detection:	Absorbance, vis 520 nm		11. Pet3Ara	—
Sample:	125 μg/mL 15 Anthocyanin standard		12. Peo3Glu	—
			13. Mal3Gal	8.76
			14. Peo3Ara	12.7
			15. Cyanidin	—
			16. Mal3Glu	—
			17. Mal3Ara	—
			18. Petunidin	—
			19. Peonidin	—
			20. Malvidin	—

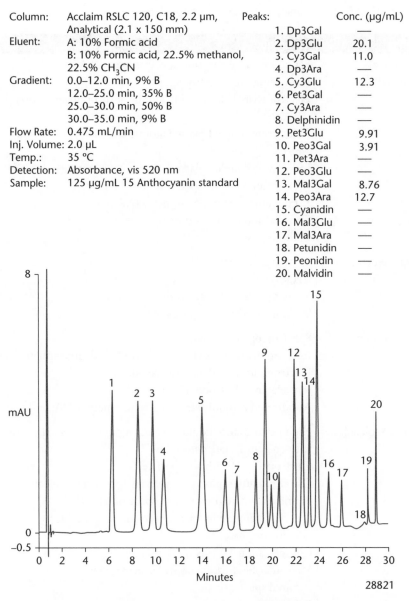

28821

Figure 4.11 Separation of anthocyanins and anthocyanidins using reversed phase HPLC (Source: Thermo Fisher Scientific. 2016. *Rapid and Sensitive Determination of Anthocyanins in Bilberries Using UHPLC*. Application Note 281. Waltham, MA: Thermo Fisher Scientific, Inc.)

. . . .

4.5 Classification of analytical techniques by analyte

A detailed description of the other techniques listed in Table 4.1 will not be given here because it is not necessary within the scope of the book, and there are numerous comprehensive treatises on the subject which can be referred to if necessary.

In practice, a more useful way of classifying analytical techniques is based upon the analyte or property to be measured, as shown in Table 4.6 and Figure 4.12. For example,

Table 4.6 Classification of analytical techniques by analyte

Analyte	Technique
Trace metals and metalloids	Flame atomic absorption spectroscopy (FAAS)
	Flame emission spectrometry
	Electrothermal atomization atomic absorption spectroscopy (ETAAS)
	Inductively coupled plasma atomic emission spectroscopy (ICP-AES)
	Inductively coupled plasma mass spectrometry (ICP-MS)
	Atomic fluorescence spectroscopy (AFS)
	X-ray fluorescence spectrometry (XRF)
	Complexation-UV-vis spectroscopy
	Cathodic voltammetry (CV)
Other elements	Total organic carbon (TOC)
	CHNS analyser
	X-ray fluorescence spectrometry (XRF)
	Scanning electron microscopy (EDXS)
Anions/cations	Spectroscopy (UV-vis)
	Ion chromatography (IC) using HPLC with appropriate detector
	Electrophoresis with appropriate detector
	Ion selective electrodes (ISE)
	Anodic and cathodic stripping voltammetry (ASV and CSV)
Organic molecules	High performance liquid chromatography (reversed phase HPLC) with an appropriate detector
	Electrophoresis with appropriate detector
	Gas chromatography (GC) with an appropriate detector
	Liquid chromatography mass spectrometry (LC-MS)
	Gas chromatography mass spectrometry (GC-MS)
	UV-vis spectroscopy
	Complexation-UV-vis spectroscopy
	Molecular Fluorescence (MF) spectroscopy
	Infrared spectroscopy
	Raman spectroscopy
	Nuclear magnetic resonance (NMR) spectroscopy
Structural techniques	X-ray diffraction (XRD)
	Thermal analysis (TA), differential thermal analysis (DTA), and differential scanning calorimetry (DSC)
	Nuclear magnetic resonance (NMR) spectroscopy
	Infrared spectroscopy (functional groups)
Field techniques	pH/Eh
	Dissolved oxygen by O_2 selective electrode
	Salinity by refractometry
	Conductivity and total dissolved solids
	Dissolved ions (NO_3^-, NO_2^-, NH_3, PO_4^{3-}, etc.)
	Metals by XRF

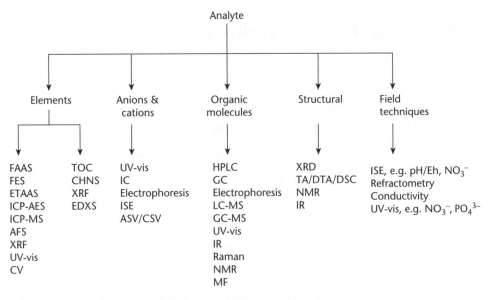

Figure 4.12 Classification of analytical techniques by analyte

if we wished to measure the levels of metallic elements, we can choose from a range of techniques including that of FAAS already described, but also X-ray fluorescence spectrometry (XRF), also based on atomic spectroscopy, and cathodic voltammetry (CV), an electrochemical technique.

It is also worth mentioning here that there are many cases where a particular material or compound has been produced or supplied and confirmation of its structure or form is all that is required. In this case, a **qualitative** analysis may provide sufficient confirmation. There are certain techniques that can be used to achieve this, such as IR spectroscopy, NMR spectroscopy, MS, UV-vis spectroscopy, XRD, and TGA. These might be employed to obtain structural information, known to be selective to the material under investigation. One example might be the preparation of an organic compound in a pharmaceutical company's laboratory. If correctly synthesized and isolated from other by-products, this main product material should possess a C=O functional group in its structure and no starting material (which possesses a C–OH group) should be present. IR analysis will allow both groups to be identified and an estimate of the purity of the product can be made.

A further category is that of **field techniques**. These are usually portable versions of laboratory instruments, such as ion selective electrodes, and hand-held spectrometers.

ACTIVITY 4.1 Choice of technique

Using Tables 4.1 and 4.6, (plus Figure 4.12), together with the details in section 4.3, identify:

1. a suitable technique to measure the carbon content of an organic sample;
2. a suitable technique to measure the carbonyl content (C=O) of a sample.

4.6 **Choosing the right technique**

In order to make the right choice of technique it is worth focussing once again on the process we used in Chapter 3 (Figure 3.1), of categorizing our sample type and identifying the property to be measured, and see how this actually links with the measurement technique. We know that we need to identify several things about the sample:

- its physical state and how it will end up after processing (i.e. as a solid, liquid, or gas);
- whether the measurement is inorganic, organic, or physical in nature;
- the type of information required, be it elemental, molecular, or structural;
- and finally, whether it is a qualitative or a quantitative measurement.

It is now possible to classify the various analytical techniques that are available based upon the analyte or property to be measured and physical state, as shown in Table 4.7.

Table 4.7 Classification of analytical techniques based upon physical state of sample and analyte together with type of analyte

Physical state		Technique	Basis of technique
Solid			
Molecular		Infrared	Vibrational
		Raman	Vibrational
		NMR	Nuclear spin
Elemental		Atomic emission	Laser ablation-ICP-AES
			Arc and Spark emission
		Atomic—mass	Laser ablation-ICP-MS
			SIMS
		X-ray fluorescence	X-ray emission using ED and WD systems
		TXRF	Thin-angle (of above)
		CHNS analyser	IR, therm. cond.
		SEM (microscopic)	Element identification, mapping, EDXS
		TEM (microscopic)	As above
	Radio-chemical	α, β, and γ spectrometry	Natural and synthetic radionuclides
		Neutron activation analysis	Gamma counting (± NAA)
Structural		X-ray diffraction	Atomic and molecular lattices
		Thermogravimetry	Measurement of changes in temperature, mass, and heat
		Differential thermal analysis	TGA/DTA
		Differential scanning calorimetry	DSC

Table 4.7 (*Continued*)

Physical state		Technique	Basis of technique
Liquid or solution			
Molecular		UV-visible	Electronic
		Molecular fluorescence	Electronic
		Infrared	Vibrational
		Raman	Vibrational
		NMR	Nuclear spin
		ESR	Electron spin
	Volatiles	Gas chromatography	With detection: FID, ECD, FPD, NP (-MS)
	Non-volatiles	High performance liquid chromatography Ion, NP, RP, GPC(SEC), Chiral	With detection: UV-vis, fluorescence, conductivity, refractive index, -MS, (with elemental tag -ICP-AES, ICP-MS)
		Electrophoresis	Various detection systems similar to above
	Biochemical	Immunochemical	ELISA (enzyme-linked immunosorbent assay)
		Enzymic	Kinetic-based
Elemental		Atomic emission	ICP-AES Flame atomic emission/photometry
		Atomic absorption	Flame atomic absorption spectrometry (FAAS)
		Atomic—mass	ICP-MS
		X-ray fluorescence	X-ray emission using ED and WD systems
		TXRF	Thin-angle (of above)
		CHNS analyser	IR, Therm. Cond.
	Volatiles	Atomic Fluorescence	AFS, Electronic
	Electro-chemical	Voltammetry	Anodic stripping, cathodic stripping, and polarography
		Potentiometry	Ion selective electrodes
		Conductometry	
		Coulometry	
		Electrogravimetry	
	Radio-chemical	α, β, and γ spectrometry	Natural and synthetic radionuclides
		Neutron activation analysis	Gamma counting ($\pm$ NAA)

Table 4.7 (*Continued*)

Physical state		Technique	Basis of technique
	Separation of elemental/ ionic species	High performance liquid chromatography Ion, chelation	With detection: UV-vis, fluorescence, conductivity, refractive index, (with elemental tag ICP-AES and ICP-MS)
	Gas		
Molecular		Molecular fluorescence	Electronic
		Infrared	Vibrational
		Raman	Vibrational
		Microwave	Rotational
	Separation of volatiles	Gas chromatography	With detection FID, ECD, FPD, NP (-MS)
Elemental	Volatiles	Atomic emission	ICP-AES
		Atomic absorption	AAS
		Atomic fluorescence	AFS
		Atomic—mass	ICP-MS
	Radio-chemical	α, β, and γ spectrometry	Natural and synthetic radionuclides, e.g. radon, UF_6, etc.

EXAMPLE 4.2

Why are my exotic fish still dying?

Consider once again the case of the exotic fish from previous chapters, where we wanted to answer the question 'Why are my exotic fish dying?' The possible scenarios identified in Chapter 3 were:

1. a liquid, water-based sample requiring an inorganic chemistry measurement, that is quantitative, of the molecular ions, i.e. nitrate, nitrite, and ammonium, in solution from the tank;

2. a liquid, water-based sample requiring an organic chemistry measurement, that is quantitative, of molecular compounds, that is, organic nitrogen-containing compounds to be measured in solution from the tank;

3. a solid sample requiring an inorganic chemistry measurement, from a mixed inorganic and organic material that is quantitative, of an element, that is, total nitrogen content to be measured in the sediment (mainly highly insoluble silicate and aluminosilicate mineral granules) from the tank which is contaminated with various organic components (fish food, fish faeces, dead organic matter, etc.).

Now, by using Tables 4.1, 4.6, and 4.7 we can see that these three scenarios can be addressed by the following three methods:

1. A quantitative measurement of inorganic molecular ions like nitrate, nitrite, and ammonium in aqueous samples may be achieved using ion selective electrodes (ISEs) and ion chromatography (IC) with a suitable detector (such as

UV absorption, conductivity, or refractive index). The technique finally chosen will be dependent upon the **selectivity** (Section 4.6.2) of that technique and its **working range** (Section 4.6.1). The separation of the three ions, nitrate, nitrite, and ammonium, using IC and their individual detection and measurement will provide greater selectivity and efficiency than the ISE. Also, three separate ISEs, each with a separate calibration, would need to be used to achieve the same measurement, and the presence of other ions could cause possible interferences.

2. A quantitative measurement of the organic molecular compounds could be performed by reversed phase-HPLC with a suitable detection system such as UV-vis or even mass spectrometry. If the nitrogen-containing compounds are volatile, then gas chromatography with a nitrogen–phosphorus (NP) detector or mass spectrometer could be used, depending on the required selectivity and working range of detection.

3. You decide by completing the next activity.

ACTIVITY 4.2 Measurement technique for nitrogen determination

Scenario 3 in Example 4.2 requires the total nitrogen content (quantitative) of the solids in the fish tank to be measured. This 'solid' was found to be present, mainly as settled material within the inert sediment at the bottom of the tank and our extended activity from Chapter 3 allowed us to identify a number of sample preparation methods to get our total nitrogen content from these solids into solution. Two of these methods involved an acid digestion, one with and one without an oxidation step. This digestion process left us with the two nitrogen-containing compounds, nitrate (oxidation of N to NO_3^-) and ammonium (no oxidation where N converts to NH_4^+) ions, respectively. Given these two possible analytes of interest, what techniques might be considered, using Tables 4.1, 4.6, (plus Figure 4.12), and 4.7 together with the details in section 4.3, to provide us with the required measurement? You will need to justify the choices made.

Other techniques are noted that might allow direct measurement of total nitrogen in the solid, as well as in the digested samples. Which techniques are these? [Hint: Consider Table 4.7 with the details given to techniques in section 4.3.]

4.6.1 Working range

Every analytical technique has a **working range** within which a quantitative measurement can be made and with a specified degree of confidence. The working range can be defined variously, but for our purpose it is between the **lower limit of detection (LOD)**, or sometimes the **limit of quantitation (LOQ)** and the **upper linear part of the calibration**. It is important to consider the working range in conjunction with the level of the analyte or property to be measured in the processed sample. The LOD and LOQ are terms which will be described in more detail later in Chapter 9, but essentially they tell you the lowest concentration of analyte that can be determined in a sample with a given degree of confidence. While the LOD allows you 'to detect' your analyte, you would use the LOQ for 'quantification' purposes. As a 'rule of thumb' only and as a guide, the value of a LOQ is approximately $3.3 \times$ LOD. The working range when noted in terms of 'orders of magnitude' can often be applied together with the LOD value to give an approximate upper concentration guide value. For example if a LOD is 1 mg L^{-1} for an analyte using a particular technique and the technique's working range is 10^2, then that analyte's upper working concentration will be approximately 100 mg L^{-1}. A term which has been used frequently in this and previous chapters, and defined in Chapter 1, is **sensitivity**, the change in signal per unit change in amount of analyte (see page 10, Problem 1.1, and Feedback on page 15). This is often wrongly used as a proxy for how well a technique can detect low concentrations of analyte. However, while sensitivity is important, it is ultimately the level of **background noise** which determines the lowest concentration which can be detected. Sensitivity and limit of detection are discussed in more detail in Chapter 5.

The working range for any given technique is constantly changing as advances are made in technology. However, once an initial selection of possible technique(s) has been made a more refined choice may be made from Table 4.8.

Table 4.8 Working range for specific techniques

Analytical technique	Normal working range/orders of magnitude	Typical limit of detection	Comments
Flame AAS	10^2	~50 µg L^{-1}	Mg = 1 µg L^{-1}; Zn = 30 µg L^{-1}; Ce = 100,000 µg L^{-1}
Flame AES	10^3	100 µg L^{-1} for Na	Sensitivity usually poor for all elements except alkali and alkaline earths
ETAAS	10^2	~0.5 µg L^{-1}	Cd = 0.05 µg L^{-1}; Zn = 1 µg L^{-1}; W ≥1000 µg L^{-1}
ICP-AES	10^5	~10 µg L^{-1}	Sr = 0.05 µg L^{-1}; Zn = 10 µg L^{-1}; Cl = 200,000 µg L^{-1} Worst metal = Cs = 3200 µg L^{-1}
ICP-MS	10^5	~0.05 µg L^{-1}	U = 0.001 µg L^{-1}; Zn = 0.05 µg L^{-1}; K = 3.1 µg L^{-1}
AFS	10^7	~0.01 µg L^{-1}	Hg = 0.001 µg L^{-1} As, Bi, Sb, and Te = 0.01 µg L^{-1}
XRF	Up to 10^6	Often around 0.1 to 1 mg kg^{-1} by solids preparation technique but extremely variable	Mn or Fe = 0.01 mg kg^{-1}; Zn = 2 to 3 mg kg^{-1} (solids technique); Light elements, e.g. Na = 1000 mg kg^{-1}
Complexation-UV-vis	10^2	~50–200 µg L^{-1}	Example: Fe-1,10 -phenanthroline = 20 µg L^{-1}
Complexation-fluorescence	10^4	~0.05 µg L^{-1}	Example: Al-lumogallion ~0.04 µg L^{-1}
Anodic stripping voltammetry	10^3–10^4	0.1–1 µg L^{-1}	LOD is dependent on plating time
Total organic carbon	~10^5	~5 µg L^{-1}	

Table 4.8 (*Continued*)

Analytical technique	Normal working range/orders of magnitude	Typical limit of detection	Comments
CHN analyser	10^3–10^4	For 2 or 3 mg sample mass, LOD = 0.05%	
UV-vis spectroscopy	10^2	~50–100 µg L^{-1}	
Molecular fluorescence	10^4–10^5	1–10 µg L^{-1}	
Ion chromatography	10^3	5–10 µg L^{-1}	
Ion selective electrodes	10^4–10^5	100–1000 µg L^{-1}	
HPLC	10^2: UV-vis 10^3: fluorescence	50–500 µg L^{-1} 0.1–10 µg L^{-1}	Almost generic Rigid aromatic compounds
GC	10^6: TCD 10^7: FID 10^6: ECD 10^5: FPD 10^5: FTD	100 µg L^{-1} 1 µg L^{-1} 0.01 µg L^{-1} 0.01 µg L^{-1} 0.01 µg L^{-1}	Any compound Any compound with C Halogenated compounds S and P compounds N and P compounds
LC-MS	10^4–10^5	~1 µg L^{-1}	
GC-MS	10^5	10–100 µg L^{-1}	
IR		Generally, at least 0.5 mg required	More required if KBr disk or Nujol mull used. Less required if IR microscope used.
Raman		1000–20,000 µg L^{-1}	Dependent on slit width and measurement time
NMR	Linear range limited to analyte solubility	~0.05 mg for ^{13}C ~0.5 mg for ^{1}H	LOD is very time dependent
Thermal techniques	e.g. TGA, DTA, etc.	10 mg in 100 g	
X-ray diffraction	10^3	~1000 mg kg^{-1} (0.1%)	
Salinity by refractometry	10^2	1‰	
Dissolved ions (portable colorimetric methods)	10^3 but different kits required	20 µg L^{-1} for PO_4^{3-} 12 µg L^{-1} NH_3 100 µg L^{-1} for NO_3^-	Example PO_4^{3-} = 0–800 µg L^{-1} or 0–4000 µg L^{-1} or 0–40,000 µg L^{-1}

Table 4.8 (*Continued*)

Analytical technique	Normal working range/orders of magnitude	Typical limit of detection	Comments
Dissolved oxygen	10^3–10^4	10 µg L^{-1}	
Total dissolved solids	10^3–10^4	1000 µg L^{-1}	
Chemical oxygen demand (COD); portable colorimetric methods	Up to 10^3 but different range kits required	Can be difficult to control below 10 mg L^{-1} for portable colorimetric systems	Example 0–150 mg L^{-1} low 0–1500 mg L^{-1} high 0–15,000 mg L^{-1} high+

4.6.2 Selectivity

Selectivity is the ability to distinguish one thing from another, in our case identifying the analyte or property and distinguishing it from everything else in the sample. The level of selectivity provides a measure of confidence in the result. If two analytes have similar properties or characteristics then each could interfere with the identification and measurement of the other unless the level of selectivity is raised to allow them to be distinguished. **Specificity** is another term that is commonly used, and can be thought of as an extended or ultimate form of selectivity where an analytical method can be used to measure the analyte or property of interest, considered suitably free from interference due to other components.

Selectivity is influenced by both sample preparation and the analytical technique used for the measurement.

Sample preparation—the level of selectivity can be increased during sample preparation by:

- chemical pre-treatment, for example, creating conditions that cause only the analyte to be free in solution while other concomitants are insoluble or immobilized or bound in the chemical sense;
- separation, for example, solid phase extraction to separate the analyte from the sample matrix.

Analytical technique—the selectivity of an instrumental technique is dependent upon:

- the characteristic being used to identify the analyte;
- the instrumental arrangement and;
- the detection system employed.

The selectivity of analytical techniques can be graded, as shown in Table 4.9.

Table 4.9 Selectivity of some techniques based upon molecular, atomic, and structural measurement systems

Technique	Selectivity	Degree of selectivity
Molecular/organic analytes		
UV-visible spectroscopy	Electronic absorbance bands around λ_{max}	Low in mixtures, linked to 'types of transition' and broad range types of organic molecule, e.g. aromatic
IR and Raman spectroscopy	Vibrational absorbance bands around λ_{max}	Selective to functional groups and bond types only, e.g. C=O etc.; low in mixtures
Molecular fluorescence spectrometry	Electronic excitation with fluorescent emission around λ_{max}	Can be more selective but not in mixtures of > 2 main components
High performance liquid chromatography	Separation of low to non-volatile mixtures of organic molecules	Increasing selectivity when tailored to property of molecule and type of detector used
HPLC as reversed phase (RP)	Lower polarity molecular species. Mobile phase is more polar than stationary phase	Refractive index < conductivity < UV-vis and IR absorbance < molecular fluorescence < mass spectrometry
HPLC as normal phase (NP)	High polarity molecular species. Mobile phase is less polar than stationary phase	As above
HPLC as chiral phase	Chirality of molecules	As above
HPLC as gel permeation (size exclusion)	Molecular size	As above
HPLC of organometallic and organometalloid molecular species	Separation of low to non-volatile mixtures of organometallic and organometalloid species by RP, NP, and ion (often aqueous) systems.	As above but also can be coupled to plasma techniques (ICP-OES and ICP-MS) when certain elemental constituents present, e.g. Fe, Mn, Zn, Cu, As, Se, Sb, etc. for very high selectivity-specific speciation measurements
Electrophoresis	Separation of organic molecules superior to HPLC	As above
Gas chromatography (GC/GLC)	Separation of volatile mixtures of organic molecules	TCD < FID < PID, ECD < NP < MS (see detectors for GC in applications section)

Table 4.9 (*Continued*)

Technique	Selectivity	Degree of selectivity
Nuclear magnetic resonance (NMR) spectroscopy	Magnetically modified energy levels of the nucleus in an external magnetic field. Nucleus dependent: ^{1}H, ^{13}C, etc.	Not selective for complex mixtures. Only selective after suitable separation techniques have been employed. Generally considered as qualitative to semi-quant. technique (see structural analysis techniques)
Molecular/inorganic analytes		
UV-visible spectroscopy	Electronic absorbance bands around λ_{max}	Can be more selective in tailored colorimetric chemistry, e.g. PO_4^{3-}, NO_3^-, NO_2^-, NH_4^+; to exercise care if simple absorbance used
IR and Raman spectroscopy	Vibrational absorbance bands around λ_{max}	Not selective unless organic-like moiety attached (CN, CO, CO_2); to exercise care if simple absorbance used
Molecular fluorescence spectrometry	Electronic excitation with fluorescent emission around λ_{max}	To exercise care if simple absorbance/emission systems used
Ion selective electrode (ISE)	Selective ion transport (e.g. NO_3^-, NH_4^+, etc.) across membranes to demonstrate a potential difference	As its name suggests, it is selective and not specific; presence of other ions in high concentration can cause substantial interference
High performance liquid chromatography (HPLC)	Separation of low to non-volatile mixtures of inorganic molecules and molecular ions	Increasing selectivity when tailored to property of molecule and type of detector used
HPLC as normal phase (NP) and as ion chromatography	Separation of high polarity and ionic molecular species (in NP mobile phase is less polar than stationary phase) and simple element ions, e.g. halogen anions; simple metallic cations/metalloid anions etc.	Refractive index < conductivity < UV-vis and IR absorbance < (molecular fluorescence and direct mass spectrometry—very limited usage) Can be linked to plasma techniques (OES and MS) when certain elemental constituents present, e.g. organometallic/organometalloids, for very high selectivity speciation measurements

Table 4.9 (*Continued*)

Technique	Selectivity	Degree of selectivity
Electrophoresis	Separation of inorganic molecules and ions—superior to HPLC	As above
Atomic/elemental analytes		
Ion selective electrode (ISE)	Selective ion transport of element ions (e.g. Cl, F, Na, H, etc.) across membranes to demonstrate a potential difference in aqueous systems	As its name suggests, it is selective and not specific. Presence of other ions in high concentration can cause substantial interference
Anodic and cathodic stripping voltammetry	The potential difference required to release (strip) an anion or cation from its 'plated' form on an electrode to solution; the anion/cation analytes having previously been plated out from the prepared sample solution	Selectivity is dependent upon the anions and cations present at the time of the analysis and the potential difference range used in the two electrochemical processes involved; care has to be exercised with this highly sensitive technique
		The techniques below are considered highly selective
Flame emission spectrometry	Specific to the analyte's atomic emission wavelengths due to electronic transitions	Element specific system but also dependent upon the selectivity of the spectrometer used for detection
Flame atomic absorption spectroscopy (FAAS)	Specific to the absorption of the analyte's atomic emission wavelengths by the analyte of interest itself; electronic transitions as a 'lock and key' system	Element specific system
Electrothermal atomization atomic absorption spectroscopy (ETAAS)	Specific to the absorption of the analyte's atomic emission wavelengths by the analyte of interest itself; electronic transitions as a 'lock and key' system	Element specific system
Inductively coupled plasma atomic emission spectroscopy (ICP-AES)	Specific to the analyte's atomic emission wavelengths due to electronic transitions from excited states	Element specific system

Table 4.9 (*Continued*)

Technique	Selectivity	Degree of selectivity
Inductively coupled plasma mass spectrometry (ICP-MS)		Element specific system when corrected for polyatomic and isobaric interferences
Atomic fluorescence spectroscopy (AFS)	Specific to the analyte's atomic emission wavelengths due to electronic transitions from excited states	Element specific system but also dependent upon the selectivity of the spectrometer used for detection
X-ray fluorescence spectrometry (XRF)	Specific to the analyte's atomic emission wavelengths due to inner-electronic transitions from created excited states	Element specific system but also dependent upon the selectivity of the spectrometer used for detection
Scanning electron microscopy (EDXS)	Specific to the analyte's atomic emission wavelengths due to inner-electronic transitions from created excited states	Element specific system but also dependent upon the selectivity of the spectrometer used for detection. Qualitative to semi-quantitative at best
Complexation-UV-vis spectroscopy	Electronic absorbance bands around λ_{max}; element to produce specific molecular species for measurement	Chemistry-tailored to element/metal/metalloid; care needs to be exercised in its use
Total organic carbon (TOC)	Based upon highly selective detectors of the oxides of the combusted compound focussing on C as CO_2	Analyte selective—specific detection
CHNS analyser	Based upon highly selective detectors of the oxides of the combusted analyte compound, focussing on C, H, N, and S as CO_2, H_2O, NO_2, and SO_2	Analyte selective—specific detection

Table 4.9 (*Continued*)

Technique	Selectivity	Degree of selectivity
Structural analysis		
X-ray diffraction (molecular and ionic/polar lattices) of solids; can be organic and inorganic	The measurement of the diffraction angle of monochromatic X-rays as they pass through crystalline lattice planes specific to a particular solid material	Provides structural lattice information for both organic molecular and polar-ionic/inorganic materials. Can provide major crystalline phases in solid mixtures with high selectivity. Semi-quantitative in many cases for complex mixtures
Thermogravimetric analysis, differential thermal analysis, differential scanning calorimetry; can be organic and inorganic	Methods based on the measurement of changes in temperature, mass, and heat transfer of solids and liquids	Not selective for unknown mixtures. May be considered partly selective for certain characterized materials on a comparative and performance basis. Provides some structural information for solids
Nuclear magnetic resonance (NMR) spectroscopy for organic molecular compounds	Magnetically modified energy levels of the nucleus in an external magnetic field. Nucleus dependent: ^1H, ^{13}C, etc.	Not selective for complex mixtures. Can be highly selective in preparative routes or after suitable separation techniques have been employed and provides structural organic chemical information. Generally considered as a qualitative to semi-quant. technique
Infrared spectroscopy of solids, liquids, and gases; can be organic and inorganic	Vibrational states of bonded atoms in a structured compound	Selective to functional groups and certain bonded atoms in compounds
Organic mass spectrometry	The mass of various fractions from the material undergoing breakdown analysis	The fragmentation pattern and mass fractions produced together with molecular ion data provide structural information

Use of this table for choice in the instrumental domain can be demonstrated by looking at the following example.

EXAMPLE 4.3

A selectivity case study of PAHs in foodstuffs

Table 4.10 shows a range of polycyclic aromatic hydrocarbons (PAHs) that can be found in the environment. Some of these may be found in foods, depending upon the exposure of the foodstuff to outside environmental factors or to certain cooking conditions. As these compounds are known to be toxic, and these toxicities are variable, the need to both identify and measure them individually can be important.

Table 4.10 Polycyclic aromatic hydrocarbons

Chemical compound	Chemical compound
Naphthalene	Benzo[*a*]anthracene
Acenaphthylene	Chrysene
Acenaphthene	Benzo[*b*]fluoranthene
Fluorene	Benzo[*k*]fluoranthene
Phenanthrene	Benzo[*a*]pyrene
Anthracene	Dibenzo[*a,h*]anthracene
Fluoranthene	Benzo[*g,h,i*]perylene
Pyrene	Indeno(1,2,3-*cd*)pyrene

The selectivity of the analytical method used to measure the PAHs can be influenced in several ways, as follows:

1. Sample preparation—there are a number of sample preparation methods which can be used to extract PAHs from solid samples into a suitable organic solvent. Solid phase extraction is then often used to separate and pre-concentrate the PAHs from the extract and other matrix components. After elution into a sample vial they would be ready for further studies. A certain amount of selectivity has been introduced during this sample preparation, but successful individual identification and measurement still depends upon the selectivity of the analytical measurement technique that is chosen.

2. Detection—we could take advantage of the fact that PAHs have absorbance in the UV-visible part of the spectrum due to electronic transitions within the molecular bonded systems. In fact, there is a common wavelength chosen for such measurements at 254 nm; however, the broad wavelength range covered by molecular absorption bands means that the PAHs will overlap in the spectrum and interfere with each other. It may be possible to improve the selectivity a little by using the technique of molecular fluorescence. This is where we use a particular wavelength to excite each analyte in sequence and look at the subsequent fluorescence (emission) signal, at each stage. However, in truth the same problem of possible overlapping bands can occur, giving rise to multiple absorbance and emission effects. This will result in an interference when there are more than two PAHs present.

3. Chromatographic separation—selectivity can be dramatically increased, however, if a separation technique is introduced. From the list of techniques available in Table 4.9, HPLC would be the preferred separation technique for solid materials of low volatility in a suitable solvent, where we are looking for molecular species.

For the relatively non-polar materials like PAHs, the column to choose would be of the **reversed phase** type (i.e. non-polar stationary phase and polar mobile phase). With suitable conditions of the mobile phase and flow rate, it is possible to separate all 16 of the PAHs, so that they are **eluted** from the column at different times, being seen as 16 separate peaks by the detector (Figure 4.13).

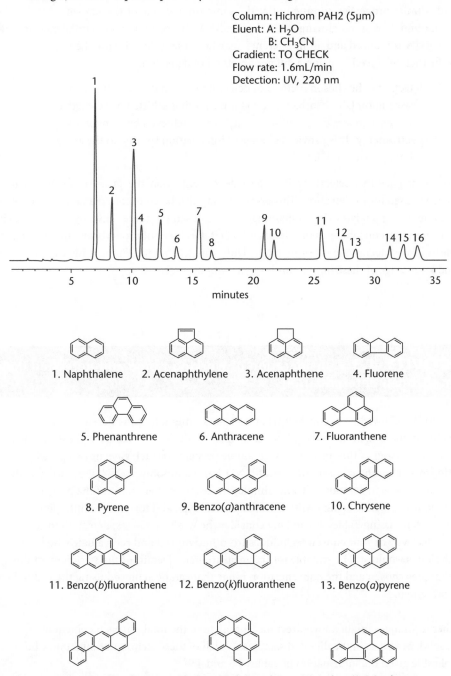

Column: Hichrom PAH2 (5μm)
Eluent: A: H_2O
　　　　B: CH_3CN
Gradient: TO CHECK
Flow rate: 1.6mL/min
Detection: UV, 220 nm

1. Naphthalene   2. Acenaphthylene   3. Acenaphthene   4. Fluorene

5. Phenanthrene   6. Anthracene   7. Fluoranthene

8. Pyrene   9. Benzo(*a*)anthracene   10. Chrysene

11. Benzo(*b*)fluoranthene   12. Benzo(*k*)fluoranthene   13. Benzo(*a*)pyrene

14. Dibenzo(*a,h*)anthracene   15. Benzo(*g,h,i*)perylene   16. Indeno(1,2,3-*cd*)pyrene

Figure 4.13 Chromatogram of 16 PAHs separated using high performance liquid chromatography with UV-vis detection (adapted with permission from: Hichrom Chromatography Columns and Supplies. 2014. *Separation of 16 EPA PAHs on Hichrom PAH2*. International Catalogue 9. Theale, UK: Hichrom Ltd)

In many cases this degree of selectivity will be satisfactory for the purposes of the analysis. However, the occasion may arise where an unknown sample type is presented to you, which may contain more PAH compounds than the 16 in this example, or other unknown components in the sample. Then further questions need to be asked if only 16 peaks are finally seen. For example, are other PAH compounds or the other compound with similar properties coming out at the same time as one of the known PAHs (i.e. a compound that is **co-eluting** with the analyte)? If so, then your confidence in the selectivity is reduced and an interference may be present. In this case the selectivity can be further enhanced by using a different method of detection:

4. Detection—the chosen method of detection used in the example so far, is based upon UV absorbance at 254 nm (or other selected wavelengths). However, a greater degree of selectivity can be achieved by using **mass spectrometry**. This provides a range of information specific to the analyte, including structural data.

So, we can see that selectivity is in part dependent upon the property of the analyte which is used for its detection. This property may also be common to a range of possible accompanying analytes (i.e. contaminants) in the sample under investigation, so each may therefore contribute an interference effect if the choice of property and the associated measurement technique is poorly made. Unfortunately, no technique is interference-free and knowledge of these limitations is of vital importance.

ACTIVITY 4.3 Consideration of working range and detection limit for nitrogen determination

Activity 4.2 allowed us to consider what techniques might be considered for the measurement of the total nitrogen content (quantitative) of the solids in the exotic fish tank. Conversion of this 'nitrogen' by a sample preparation step gave us two possible nitrogen-containing compounds, nitrate (NO_3^-) and ammonium (NH_4^+) ions to actually determine, now in solution. Various simple choices were then made of the possible techniques available, being justified on the general type and form of the 'nitrogen' now present. Using Tables 4.8 and 4.9 consider the working ranges, detection limits, and selectivity of your chosen techniques from Activity 4.2 and re-evaluate which techniques would still be suitable for the measurement. Details of the selectivity of the techniques chosen and the concentration range that can be achieved using them should be presented.

Other techniques might allow direct measurement of the total nitrogen content of the solid, as well as in the digested samples. Which of these techniques might now be applicable given the information in Tables 4.8 and 4.9?

ACTIVITY 4.4 Which analytical technique?

Explain, giving reasons, which of the analytical techniques given in Table 4.11 you would use for the following analyte/sample combinations:

(a) high levels of arsenic in mine tailings;

(b) pH in river water;

(c) individual volatile organic compounds in workplace air;

(d) low levels of arsenic in tuna fish;

(e) total polyaromatic hydrocarbons in an extract of smoke particles.

[Hint: Make your initial selection using Table 4.11 below along with Table 4.6 (and Figure 4.12), and then use Tables 4.8 and 4.9 to refine this based on the required working range and selectivity.]

Table 4.11 Analytical techniques

High performance liquid chromatography with UV-vis detection	Gas chromatography mass spectrometry	ICP-AES
ICP-MS	Ion selective electrode	UV-vis spectroscopy
Molecular fluorescence	CHN analyser	Atomic fluorescence

NB: It may be possible to use more than one technique to measure the analyte of interest (AOI).

ACTIVITY 4.5 Which analytical technique for the determination of zinc in sediment?

You have just prepared extract solutions from a river sediment by gently digesting samples in two different extractants. These soft digestion solutions are designed to mimic and extract that which might become readily 'available', that is, released, to aquatic organisms by the sediment over time. The two digestants are dilute hydrochloric acid (0.1 M) and dilute nitric acid (1.0 M). There is a limit for the 'available' Zn from sediments in some countries and in Canada the limit is 123 mg Zn per kg of sediment. Given that you end up with two digestants with different amounts of Zn in them and that the level in the 0.1 M HCl solution was estimated to be 0.1 mg L^{-1} and that in the 1.0 M HNO_3 solution was estimated to be 100 mg L^{-1}, which instrumental techniques would you consider for the measurements? [Hint: Use Tables 4.6 to 4.9 to guide you, together with the relevant sections covering more detailed descriptions of the choices available.]

Construct a table, as shown, which allows the chosen instrumental techniques and their important operating parameters to be compared. These are to include: the linear working range, possible limit of detection and limit of quantitation, the selectivity, and possible interferences. [Hint: Consider the information given in Section 4.6.1 and Tables 4.8 and 4.9.]

→

→

	Linear working range/mg L^{-1}	Limit of detection/ mg L^{-1}	Limit of quantitation/ mg L^{-1}	Selectivity and possible interferences
Method A Method B Method C				

ACTIVITY 4.6 Which analytical technique for the determination of dichlorvos and phosphorus?

In Chapter 3 we identified the categories for the pesticide dichlorvos in a food crop, noting that this pesticide must not be above the maximum allowable concentration. We also identified the categories for the levels of phosphorus in a soil sample from a field in order to calculate how much fertilizer to add. Having identified the categories, we then considered the various preparative techniques required to extract both analytes from their sample types, to allow their measurement to be made.

Which instrumental methods would you consider for these two measurements and with the minimum of interferences?

Construct a table, as shown below, which allows the chosen instrumental method and their important operating parameters to be compared. These are to include: the linear working range, possible limit of detection and limit of quantitation, the selectivity, and possible interferences associated with these two techniques. More than one method can be considered for each. [Hint: Use the tables and sections employed previously for the other activities as a guide. You might also consider again the terms 'pesticide residue' and 'dichlorvos' in any additional searches you make; e.g. https://www.ams.usda.gov/sites/default/files/media/PDP%20Abstract%20for%20Analytical%20Methods.pdf]

	Linear working range/mg L^{-1}	Limit of detection/mg L^{-1}	Limit of quantitation/ mg L^{-1}	Selectivity and possible interferences
Method A Dichlorvos				
Method B Dichlorvos				
Method C Phosphorus				
Method D Phosphorus				

PROBLEM 4.1

Case studies in sample preparation and analysis

From the entire list shown, which was introduced to you in Problem 1.2, choose one of the analytical case study problems to solve. Using the techniques discussed so far, prepare an analysis report based on the chosen case study identifying the following points:

(i)   definition of the problem;

(ii)   sampling and the sampling plan;

(iii) sample preparation;

(iv)   choice of measurement technique.

The case studies to choose from are as follows:

1. A sewage problem: urea in a coastal seawater sample (EA)

2. A food problem: mercury in tuna (FSA)

3. A nutrition problem: Se in our diet (FSA)

4. A metals impurity problem: sulfur in pig iron/blast furnace (industrial)

5. An air pollution problem: H_2S emission from a flue stack at a coal power station (energy generation)

6. A clinical problem: high sodium in a blood sample (hospital trial with new drug)

7. A pharmaceutical problem: steroids (prednisolone) level in a tablet, contaminant

8. A metals production problem: manganese levels in a stainless steel sample

9. A mining problem: arsenic levels in an ore sample (industrial and EA)

10. A soil problem: a pesticide (Gamazene/HCCH) residing in soil at a farm

11. A sediment problem: tin levels in a marina-dredged sample (marine and EA)

12. A biological plant problem: a copper-based fungicide in tea leaves (agricultural industry and FSA)

13. A petroleum problem: Pb in a sample of petrol from abroad (Customs and Excise, now HMRC)

14. A health products problem: silicon in a toothpaste sample (industrial)

15. An occupational accident problem: boron-containing dust in air at a factory after a fire in an $LiBO_4$ flux tank (occupational health, HSE)

16. A nuclear monitoring problem: uranium oxide in a fuel rod

17. An environmental health problem: platinum group metals from road dust in a city centre

18. A plastics problem: plasticizer levels in a plastics sample (PVC manufacturer)

19. An occupational hazard problem: Pb fume levels above a float-soldering bath (occupational health, HSE)

20. A waste problem: levels of NH_3/NH_4^+ in a landfill leachate (WWT plant problem)

21. Polychlorinated biphenyls (PCBs): the presence of PCBs in contaminated ground from beneath a disused electrical sub-station (EA)

22. The level of $CaCO_3$ (a filler) in a loaded plastic: the quantity of filler affects the properties and is a quality control problem (industrial)

23. Arsenic in a paint sample: a 200-year-old stately home with green coloured main hall walls believed to have been painted with copper arsenate

24. The spillage of milk from a dairy farm to a local river: the practice of hosing-out of the cow's milking parlour each day and collection of all wash-waters in an adjacent lagoon is suspected (DEFRA and EA)

25. To determine a measure of the bioavailability of lithium, from the compound lithium carbonate, as used to treat sufferers from bipolar disorder (MRC)

26. Food additive E110 (Sunset Yellow) is suspected as being the colouring agent, present in high concentration in a child's confectionary but not disclosed (FSA and TS)

27. Levels of silver (Ag) in a fabric: silver compounds are used as a biocide to treat fabrics which may have to be worn in harsh environmental conditions (TS and MoD)

28. The levels of chloride and chlorine (free) present in a public swimming pool (Environmental Health and PA)

29. An occupational hazard problem: mercury fume/airborne levels within a dentist's surgery (Occupational/Environmental Health/HSE)

30. The levels of certain PAHs (anthracene etc.) in cigarette smoke and the effectiveness of filters (TS and PA)

31. The levels of arsenic and cadmium in allotment soil used to grow edible produce (EA and PA)

32. Bisphenol A in the diet: levels from polycarbonate (PC) baby bottles as the most prominent role of exposure for infants, and/or canned food for adults and teenagers (FSA and PA)

33. The identification and measurement of carbon monoxide levels from a suspected faulty gas-fired heater (Metropolitan Police/Crown Prosecution/HSE)

34. The measurement of levels of antimony in fruit juices and the PET bottles that contain them, as sold to the public (HSE, FSA, and PA)

35. The determination of polybrominated diphenyl ethers (PBDE; flame retardants) in household furnishings and dust, use of which, for some, has now been banned (HSE, TS, and PA)

36. The levels of lead, cadmium, and arsenic in a 'dolomite' sample, used in the preparation of Ca/Mg food supplement pills (FSA)

37. The determination of a vitamin (A or C or D or E) levels in a medicinal preparation for the treatment of patients with inflammatory bowel disease (MHA)

38. The measurement of silicon and silicone levels in blood obtained from humans who have undergone implantation with silicone-gel-containing prostheses (HSE)

39. The determination of aluminium and aluminosilicate levels in the preparation of gloss paper (QC/QA)

40. The determination of potentially toxic element contamination in macroalgae (seaweeds) consumed by 'foraging' members of the public for personal or livestock consumption (DEFRA, FSA, and PA)

41. The determination of pesticide and growth-modifier residues on fruit and vegetables, e.g. Alar on apples (FSA, PA, and DEFRA)

42. The determination of iodine/iodide levels in macroalgae (seaweeds) consumed by 'foraging' members of the public for personal or livestock consumption (DEFRA, FSA, and PA)

43. The measurement of tributyltin, an anti-fouling agent added to paints and surface preparations of hulls of ships, from a marina-based seawater or sediment sample (EA, TS, and PA)

Feedback on activities and problems

 FEEDBACK ON ACTIVITY 4.1

Choice of technique
Using Tables 4.1 and 4.6, identify:

1. *A suitable technique to measure the carbon content of an organic sample*

    The tables indicate that suitable techniques available to measure the carbon content, as an element of an organic sample, could include:

    • a total organic carbon (TOC) analyser;

    • a CHN (carbon–hydrogen–nitrogen) analyser.

2. *A suitable technique to measure the carbonyl content (C=O) of a sample*

    The tables indicate that suitable techniques available to measure the carbonyl content (C=O) of a sample could include:

    • FT-IR spectroscopy;

    • Raman spectroscopy;

    • NMR spectroscopy;

    • (UV-vis spectroscopy).

 FEEDBACK ON ACTIVITY 4.2

Measurement technique for nitrogen determination
Given these two possible analytes of interest (NO_3^- and NH_4^+), what techniques might be considered, using Tables 4.1, 4.6, and 4.7, to provide us with the required measurement. You will need to justify the choices made.

The samples containing the nitrogen compounds are now in solution and both can be measured by: ISE, visible spectroscopy, and/or ion chromatography with suitable detector. The first two techniques are relatively selective. These include using a nitrate or ammonium ion selective electrode (Nernst-like EMF measurement) or using chemical colorimetry of these ions with visible spectroscopy and the Beer–Lambert law to quantify. Ion chromatography would require an anion and a cation column, allow greater selectivity through separation and with suitable detector the measurement would allow greater specificity. Calibration is of course required in all cases, using known quantities of the specific analyte.

Other techniques might allow direct measurement of total nitrogen in the solid, as well as in the digested samples. Which techniques are these? Two solids techniques to consider might be a CHN analyser or possibly X-ray fluorescence spectrometry.

 FEEDBACK ON ACTIVITY 4.3

Consideration of working range and detection limit for nitrogen determination

Various simple choices were made of the possible techniques available, being justified on the general type and form of the 'nitrogen' now present. Using Tables 4.8 and 4.9, consider the working ranges, detection limits, and selectivity of your chosen techniques from Activity 4.2 and re-evaluate which techniques would still be suitable for the measurement. Details of the selectivity of the techniques chosen and the concentration range that can be achieved using them should be presented.

Technique	Conc. range	LOD	Selectivity
ISE NO_3^-	Can measure broad range (10^{4+}) from sub mg L^{-1} to 100's mg L^{-1}	~100+ µg L^{-1}	Ion selective but not specific—other ions can interfere if high conc.
ISE NH_4^+	Can measure broad range (10^{4+}) from sub mg L^{-1} to 100's mg L^{-1}	~100+ µg L^{-1}	Ion selective but not specific—other ions can interfere if high conc.
UV and visible spectroscopy NO_3^-	Can measure narrow range (10^{2+}) from sub mg L^{-1} to 10's mg L^{-1}	~100+ µg L^{-1}	Selective but not specific—other analytes can interfere
Visible spectroscopy NH_4^+	Can measure narrow range (10^{2+}) from sub mg L^{-1} to 10's mg L^{-1}	~100+ µg L^{-1}	Ion selective but not specific—other analytes can interfere
IC NO_3^-	Can measure rel. broader range (10^3) from 10's µg L^{-1} to 10's mg L^{-1}	~5–10 µg L^{-1}	See bullet points below
IC NH_4^+	Can measure rel. broader range (10^3) from sub mg L^{-1} to 10's mg L^{-1}	~5–10 µg L^{-1}	See bullet points below

- Ion chromatography (IC) would require an anion and a cation column but would allow greater selectivity through separation and with suitable detector the measurement would allow greater specificity from interferences.

- Ion selective electrode (ISE) measurements are quick and easy to perform and could provide a first stage value for gauging levels for further measurement techniques if required.

- Any of the techniques shown can contribute a usable value under suitable conditions but concomitants from the matrix and the limits of detection mean that ISE may not always give the required confidence.

The other techniques that might allow direct measurement of the total nitrogen content of the solid, as well as in the digested samples, are CHN analyser and X-ray fluorescence.

Technique	Conc. range	LOD	Selectivity
CHN analyser	Can measure rel. broader range (10^3–10^4) based upon absolute mass	Only 2 or 3 mg able to be used at a time and LOD only 0.05% of mass taken (1:2000)	Quite good selectivity but restricted on its overall sensitivity
X-ray fluorescence	Variable but can be broad range (10^{5+})	Can be down to ~0.1 mg/kg but not for 'N'—light element problem	Quite good selectivity but restricted on its overall sensitivity for light elements

Direct solids analysis using the techniques above would be limited in their confidence and range.

 FEEDBACK ON ACTIVITY 4.4

Which analytical technique?

(a) High levels of arsenic in mine tailings can be measured after digestion of the ore tailings and then in solution by ICP-AES. This is an element-specific technique with a broad calibration range. Atomic fluorescence spectrometry, like ICP-MS, is likely to be too sensitive so that dilution to bring the sample solution within range would multiply errors from this procedure. Also, interference of the chemistry used to produce the volatile species often used for this technique could cause a problem.

(b) The pH in river water can be measured directly using an ion selective electrode, specifically a pH electrode which is selective to H_3O^+ ions.

(c) Individual volatile organic compounds in workplace air can be measured using gas chromatography with a mass spectrometric detector. The term 'individual' requires that a separation technique be used and the volatile term indicates the ease of achieving the molecular species in the vapour/gas phase, suited to gas chromatography. The mass spectrometric detector provides mass detection of the M^+ ion and, if required, structural information on each for confirmatory identification. Volatile compounds can be captured where necessary using solvent absorbers or trapping systems outlined in Chapter 3.

(d) Low levels of arsenic in tuna fish can be measured using the element-specific techniques of ICP-MS and atomic fluorescence spectrometry. The low levels present in the original sample are further diluted by the acid digestion technique (e.g. nitric acid + hydrogen peroxide on freeze-dried and ground samples), therefore requiring instrumentation with high sensitivity and low limits of quantitation.

(e) Total polyaromatic hydrocarbons in an extract of smoke particles (e.g. solvent extraction on cigarette filters) can be measured using molecular fluorescence (MF) and possibly low resolution HPLC with UV-vis detection (HPLC-UV). Here the term 'total' means that the analytes particular molecular class or type is determined, in this case the PAHs. Because of their structure and bonding, PAHs can be detected with selective techniques such as MF and low resolution HPLC-UV, taking advantage of their ability to detect particular molecular electronic transitions.

FEEDBACK ON ACTIVITY 4.5

Which analytical technique for the determination of zinc in sediment?

Construct a table, as shown, which allows the chosen instrumental techniques and their important operating parameters to be compared. These are to include: the linear working range, possible limit of detection and limit of quantitation, the selectivity, and possible interferences.

In this case we should consider element-specific techniques which take advantage of atomic electronic transitions in atoms; in this case, zinc atoms (as ions) in solution (see Table 4.8, page 96). Two sample extract solutions were prepared with Zn at ~0.1 mg L^{-1} (~100 µg L^{-1}) and ~100 mg L^{-1}. Table 4.9 shows five or six techniques of possible use on a general approach. However, not all would be selected because of various factors. It is noted, for example, that Zn by flame atomic absorption spectrometry (FAAS) would be linear up to ~3 mg L^{-1} and its LOD might be ~30 µg L^{-1} with its LOQ ~100–150 µg L^{-1}. The lower concentration Zn extract solution to be measured would be too close to the LOQ for FAAS while the upper concentration extract solution would require further dilution by ~100 to bring it within the linear range. While the latter is possible, this means further errors will be introduced, compared with those techniques that could measure either or both extracts directly.

As with most practical decisions, they are made with what is available. The choices shown below still have some compromises to consider but will achieve the required measurements under suitable conditions.

	Linear working range/ mg L^{-1}	Limit of detection/ mg L^{-1}	Limit of quantitation/ mg L^{-1}	Selectivity and possible interferences
Method A ICP-AES	>500	~0.01	~0.03	Highly selective line emission from atoms and ions in the plasma Interferences not generally spectroscopic but mass matrix/transport based
Method B ICP-MS	>100	0.00005	~0.0002	Good selectivity based upon isotopic mass Possible interference based upon polyatomic ion formation
Method C X-ray fluorescence	>1000	~0.1	~0.3	Good selectivity based upon X-ray emission from atoms Selectivity dependent upon wavelength (WD) or energy (ED) dispersive detection Can suffer matrix absorption effects but less so in solution

? **FEEDBACK ON ACTIVITY 4.6**

Which analytical technique for the determination of dichlorvos and phosphorus?

Construct a table, as shown below, which allows the chosen instrumental method and their important operating parameters to be compared. These are to include: the linear working range, possible limit of detection and limit of quantitation, the selectivity, and possible interferences associated with these two techniques. More than one method can be considered for each.

Considering Tables 4.8 and 4.9, and for Method A, Table 4.4, together with the identification classes of the analytes and preparation methodologies discussed, then the table below is one set of possible solutions. It must be remembered that these analytes are in a particular phase and concentration after preparation. Dichlorvos is an organic molecular species containing both chlorine and phosphorus in an organic extract and is a volatile liquid (b.p. 74°C), while phosphorus is in an aqueous acidic solution, present as the molecular ion 'phosphate' and as such may be targeted as the element or as the phosphate ion.

In foods, dichlorvos limits are based upon daily consumption and selected levels in specific food types. For example, the US EPA acute toxicity for this compound is based upon a 70 kg male, which is 700 μg day^{-1} or 10 μg kg^{-1} body weight day^{-1}. The US EPA has established maximum permissible levels of dichlorvos in various food products ranging from 0.02 to 2 ppm. Samples from the food supply are regularly tested for dichlorvos. The analyst therefore often has to consider a range of options based upon intake and worst-case scenarios, and how this would result in a selected range of levels after sample preparation for measurement, if a level is not specified.

	Linear working range/mg L^{-1}	Limit of detection/mg L^{-1}	Limit of quantitation/mg L^{-1}	Selectivity and possible interferences
Method A for dichlorvos: GC with EC, NP, or MS detectors	EC up to ~10 NP up to ~10 MS up to 100	~10^{-6} ~10^{-5} ~10^{-2}	~10^{-5} ~10^{-4} ~10^{-2}	See Tables 4.4, 4.8, and 4.9 for details
Method B for dichlorvos: HPLC-UV	Up to ~10	~5×10^{-2}	~10^{-1}	See Tables 4.5, 4.8, and 4.9 for details
Method C for phosphorus: ICP-AES ('total' phosphorus measured)	Up to ~500	~2×10^{-2}	~6×10^{-2}	See Tables 4.8 and 4.9 and Section 4.4.1.2 for details (Note: all forms of phosphorus measured)
Method D for phosphorus: UV-vis spectrophotometry (phospho-molybdate blue reaction in visible)	Up to ~10	~5×10^{-2}	~10^{-1}	See Tables 4.8 and 4.9 and Section 4.4.2.1 for details (Note: phosphorus as PO_4^{3-} is measured; arsenic and silicon can interfere)

FEEDBACK ON PROBLEM 4.1

Case studies in sample preparation and analysis

There are various combinations of sampling, sample preparation, and measurement that can be levelled at each of the case studies presented and hence various successful solutions depending upon the extent of each scenario set. However, the requirement that the technique is '**fit for purpose**' encapsulates the individual requirements of representative sampling and sample preparation that maintain the integrity of the analyte while the measurement technique is suitably sensitive, selective, and free from interference.

A good starting point for any method development is to interrogate the literature for regulatory analytical procedures active at the time. However, there are many sample types and analytes that are yet to be regulated or even monitored and you will need to have a good knowledge of the six steps in the analytical approach and what makes it fit for purpose. The scientific literature presented by governing, protective bodies around the world, including the WHO (World Health Organization), the OECD, and those countries in Europe, the Americas, Australasia, and Asia, each identify standard operating procedures (SOPs) for many analytical tasks. Scientific journals also publish peer-reviewed papers containing details of food, drugs, water, medical, and environmental analyses for analytes of interest, as part of an ever increasing and hopefully improving system of knowledge for life-management. One such document[4] from the New Jersey Department of Environmental Protection is provided for public scrutiny. This is the *Field Sampling Procedures Manual* which covers all aspects of the sampling approach undertaken and the level of detail can be assessed and adjusted accordingly.

As an example, let us look at one case study in some detail, and address the problem as set out.

The levels of arsenic and cadmium in allotment soil used to grow edible produce

Definition of the problem

Consider the problem of sampling an actual field site, as we have in previous chapters. If we extend this scenario to include, for example, the requirement to representatively sample and prepare our samples, for measurement of arsenic and cadmium in an allotment soil used to grow edible produce then we need to understand that the reason for the analysis includes regulatory levels within the Environmental Quality Standards set, which are:

Total arsenic and cadmium levels should not exceed 43 mg/kg and 1.8 mg/kg, respectively.[5]

---

[4] New Jersey Department of Environmental Protection. 2011. *Field Sampling Procedures Manual*. Available at: http://www.nj.gov/dep/srp/guidance/fspm/.

[5] Environment Agency. 2009. Reports SC050021/SGV, SC050021/SR3, SC050021/SR2, SC050021/SR TOX1, and SC050021.

This is our starting point. These levels will help us to frame the procedures later on.

Using our analytical choices flow chart (see Figure 3.1), for this scenario, we are able to define our wet soil system as:

A **solid sample** (soil components) with associated **liquid** (water) – two phases are present. Both phases can contain the analytes of interest (insoluble and soluble forms) and will contribute to the overall final value, which is required. A suitable drying process should leave any dissolved material with the sample which can be measured.

We are interested in the total arsenic and cadmium levels present in the dried soil sample. While these analytes can be present in both **inorganic and organic** forms within the, now dried, soil sample we are interested only in their total available **elemental** contribution and not their molecular or structural composition. This helps to inform us which digestion methods and which instrumental measurement techniques we could use.

Finally, we need to know their individual levels, with suitable confidence, in the dried soil to compare with the soil guideline values. We are therefore undertaking a fully quantitative measurement of both.

Further questions identify that we have an allotment area of 1 ha (10,000 m²) which has 100 allotments, each 10 m × 10 m in size. The crops grown have a maximum root depth of 30 cm.

Sampling and the sampling plan

In this example we will take into account all of the following:

- You are sampling from 100 allotments within 10,000 m² (1 ha) of land.
- A representative sampling approach should be taken.
- Assume that you will sample twice during the year to coincide with two crop periods where the allotment soil has been prepared for growth, to provide a temporal comparison.
- Assume inclusion of a known 'hot spot' defined at the 95% confidence level for the minimum number of samples.

For a 1 ha site, the minimum number of sampling points for 95% confidence is 21, with a hot spot diameter of 25.7 m, calculated using the method described in Chapter 2:

1. Determine or estimate the radius of the smallest hot spot, r (in metres), that needs to be detected.

2. Calculate the size of the grid, G, using equation (2.1):

$$G = \frac{r}{0.59}$$

3. Determine the number of sampling points required, n, for an overall size of sampling area, A (in m²) using equation (2.2):

$$n = \frac{A}{G^2}$$

However, also consider that a more logical division of 100 allotments would be into 25 sites, to give 25 samples. This reduces your hot spot to 23.6 m diameter. To reduce any systematic bias from this number, you could consider a systematic herringbone sampling array, unaligned with random start, as shown below:

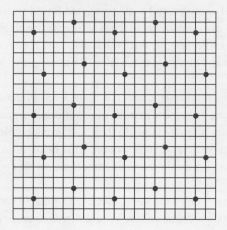

This means that individually identified samples are taken from within a 20 m × 20 m array (4 allotments, 2 × 10 m by 2 × 10 m). Using a 4 cm diameter corer to a depth of 30 cm would provide a ~0.6 kg sample from each, based upon soil density of 1.6 g cm^{-3} (and $\pi d^2 L/4$, where $L = 30$ cm and $d = 4$ cm). For comparison, a 0.94 kg sample is obtained if the corer is 5 cm in diameter.

An alternative is that you could try taking 25 samples but each site is made up of four composite sub-samples to give a total sample size of 2.5 kg. From the four allotments in each 20 m × 20 m grid, each allotment supplies ~0.6 kg and the four allotments together supply the total ~2.5 kg which are from the 4 × 0.6 kg thoroughly mixed samples, where each ~0.6 kg is taken from, say, the centre of each 10 m × 10 m allotment.

Mixing may be undertaken using a spinning riffler, or similar suitably efficient process, and then sub-sampling can be performed on this composite sample by taking 20–25% by mass.

You will then end up with 25 individually identified ~0.6 kg samples to process in the laboratory. Remember to identify the samples to indicate the location and allotments to which they relate.

The samples are to be collected in containers that maintain their integrity.

Sample preparation

In order to sub-sample and obtain test samples for digesting and/or extraction of the analytes, the samples are best processed in their dry state, and therefore a drying process should be undertaken. The analytes, As and Cd, are not considered volatile under 105°C so their drying can be accelerated in a suitable oven. The samples will need to be ground and sieved to allow multiple representative test samples of ~1 g to be taken from each ~0.6 kg sample. Hence, if all the ~0.6 kg sample is processed to provide up to 50 or 60 g (approximately 10%), of ground and sieved sample for further sub-sampling, this would suffice. The grinding technique chosen should be one which avoids contamination, using grinding media with a hardness greater than the components in the soil. Generally, for soils and sediments, one would consider grinding techniques and materials of >7 on the MOH scale. Some ball mills can both gently grind and trundle to dis-aggregate the soil components and allow them through a 180–250 μm sieve (preferably a plastic sieve). It is usual to grind a small quantity of the mixed soil in the mill and then discard it first, to coat the grinding elements and reduce contamination from the pre-cleaned mill components.

The regulation states that allotment soil should not provide more than 43 mg As or 1.8 mg Cd per kg of the dried soil. Methods for the extraction of the analyte are many and varied and, in this particular case we are mostly interested in the 'available' As and Cd, that is, available biologically for uptake into plants and organisms, rather than bound up in minerals. A relatively simple acid digest can mimic the release of available analyte so total digestion of the complete soil is not necessary or required. Some SOPs are written specifically for this, one of which recommends *aqua regia* (1 part concentrated nitric acid plus 3 parts concentrated hydrochloric acid) to extract the analyte from a known mass of soil, under reflux for 2 hours. At the end of this period the extract is cooled, filtered, if required, through a suitable pre-washed, inert filter paper. The residue is left and the filter paper washed with 1–2% nitric acid, and the extract is made up to a known volume. To accompany this process, a number of procedural blanks are prepared in exactly the same way. To allow for a measure of reproducibility, each sample and its equivalents should be processed in at least triplicate.

Choice of measurement technique

You would consider choosing the technique based upon:

- selectivity—multi-element atomic spectrometry; specific wavelengths with highest sensitivity and fewer interferences;
- sensitivity and working range—covering, ideally, 4–5 orders of magnitude with greatest sensitivity, sufficient to determine the analytes in their diluted state.

If As and Cd are extracted from 1 g samples under reflux with say 20 mL of *aqua regia*, then the extract containing the analytes can be made up to a known accurate volume, as stated in the sample preparation section. As filtering and washing of the extracts is required, then a suitable volume to make up to would be possibly 50.0 mL or 100.0 mL in a volumetric flask. If we take the latter volume (to reduce the effects of concentrated *aqua regia* being present) then this is a 1:100 dilution. If the samples were at or just under the regulatory limits, then their concentrations in the diluted extracts would be As = 0.43 mg L^{-1} and Cd = 0.018 mg L^{-1}. From looking at Table 4.8 we can see that it is possible to use ICP-AES for As but Cd may begin to approach the LOQ of 5–10 μg L^{-1}. It may therefore be prudent to consider ICP-MS to provide two orders of magnitude lower detection, if the analytes are present below the 10–20 mg L^{-1} level.

Calibration and quantitation

Learning outcomes

Once you have read through this chapter, worked through the activities and problems, and reviewed the feedback, you should be able to:

1. prepare a series of working standards to be used for instrument calibration;
2. plot a calibration curve and fit a linear trend line;
3. use a calibration curve to calculate the analyte concentration in a sample;
4. distinguish between the different sources of errors that can occur in instrumental analysis.

5.1 Introduction

This chapter explores the actuality of making an analytical measurement. As we have seen already, and which you will also discover in later chapters, there is far more to performing accurate analytical measurements than just walking into the lab and 'doing an analysis'.

5.2 Qualitative or quantitative analysis?

In this book we are primarily concerned with quantitative analysis, the measurement of the amount of an analyte in a sample. However, it is worth bearing in mind that in order to perform accurate quantitative analysis it is also necessary to be confident that you are measuring the right thing. Consequently, some form of **qualitative analysis**, normally referred to as **selectivity**, is almost always a necessary pre-condition.

Qualitative analysis concerns the **identification** of an analyte rather than the measurement of its quantity. For example, mass spectrometry can be used to identify controlled drugs in body fluids for forensic purposes. This is illustrated in Figure 5.1, which shows the mass spectrum of morphine, comprised of fragments of the molecule which have different masses,[1]

---

[1] More correctly, the spectrum shows mass-to-charge ratio (m/z) rather than mass but, because in this case the fragments usually carry only a single positive charge, this is numerically equal to mass.

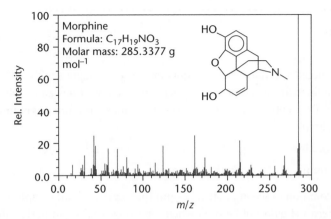

Figure 5.1 Electron ionization mass spectrum of morphine (data from NIST Standard Reference Database 69: NIST Chemistry WebBook)

this fragmentation pattern being characteristic of morphine. In fact, if there are no interfering compounds present the most intense mass fragment, known as the **base peak**, is often sufficient for identification.

Q: What m/z does the base peak for morphine have in Figure 5.1?

A: The base peak is at approximately 285 m/z, which is the same as the molar mass of morphine, so in this case the base peak represents a singly charged morphine molecule. The other peaks at slightly higher and lower mass (either side of the main ~285 m/z peak) are due to the presence of numerous naturally occurring isotopes of the elements which make up the morphine molecule.

So, given that we are confident in our ability to measure the correct analyte, in this case morphine:

Q: How might you perform a quantitative analysis of morphine using the mass spectrum shown in Figure 5.1?

A: You could measure the height of the base peak at 285 m/z, which is in turn related to the number of morphine molecules present in the original sample.

Unfortunately, the relationship is generally not straightforward, so a process of **calibration** must be performed to establish exactly what the relationship is. Calibration will be described in detail in this chapter.

Historically, quantitative analysis has been categorized according to the concentration of analyte in the sample as follows:

Major	1–100%
Minor	0.01–1%
Trace	0.1–100 mg kg^{-1}
Microtrace	0.1–100 μg kg^{-1}
Nanotrace	0.1–100 ng kg^{-1}

Note that the examples given above illustrate concentrations in solid samples, but they apply equally to liquid samples.

5.3 Units

While we have already used a variety of units in previous chapters without explanation, it is worth revisiting them at this point to establish certain conventions and apply them correctly and consistently. Concentration can be expressed as a **mass fraction** or **volume fraction** as in the following examples:

| Volume fraction: | $mg\,L^{-1}$ | $\mu g\,L^{-1}$ | $mol\,L^{-1}$ | etc. |
| Mass fraction: | $mg\,kg^{-1}$ | $\mu g\,kg^{-1}$ | $mol\,kg^{-1}$ | etc. |

It is extremely common to see units such as **ppm** (parts per million) or **ppb** (parts per billion) quoted. These terms should not be used when reporting results and performing calculations because they are ambiguous and give no indication of whether the concentration unit is a volume fraction or mass fraction. (Having said that, you will probably end up using these terms in general conversation and when talking to other analysts!) Another confusing aspect of this subject is that you will often see units quoted in a variety of different formats derived from SI units. The most commonly used mass and volume fractions are shown in Table 5.1.

Another aspect of terminology which you need to master is the prefix commonly used to express concentrations in trace, microtrace, and nanotrace analysis, given in Table 5.2.

Throughout this book we have used **mL or L for volume** and **g or kg for mass**. In addition, sometimes prefixes are used to indicate smaller quantities, for example, μL for $10^{-6}\,L$ or ng for $10^{-9}\,g$. This may not be what you have been taught in school (e.g. dm^{-3} is used instead of L^{-1}), but this is in line with the kind of units you will commonly encounter in analytical chemistry.

Table 5.1 Common usage and the correct forms of concentration units

Common usage		Volume fraction				Mass fraction	
parts per million	ppm	$\mu g\,mL^{-1}$	$\mu g\,cm^{-3}$	$mg\,dm^{-3}$	$mg\,L^{-1}$	$\mu g\,g^{-1}$	$mg\,kg^{-1}$
parts per billion	ppb	$ng\,mL^{-1}$	$ng\,cm^{-3}$	$\mu g\,dm^{-3}$	$\mu g\,L^{-1}$	$ng\,g^{-1}$	$\mu g\,kg^{-1}$
parts per trillion	ppt	$pg\,mL^{-1}$	$pg\,cm^{-3}$	$ng\,dm^{-3}$	$ng\,L^{-1}$	$pg\,g^{-1}$	$ng\,kg^{-1}$
parts per quadrillion	ppq	$fg\,mL^{-1}$	$fg\,cm^{-3}$	$pg\,dm^{-3}$	$pg\,L^{-1}$	$fg\,g^{-1}$	$pg\,kg^{-1}$

Table 5.2 Commonly used concentration prefixes

Prefix	milli	micro	nano	pico	femto	atto	zepto
Symbol	m	μ	n	p	f	a	z
Meaning	10^{-3}	10^{-6}	10^{-9}	10^{-12}	10^{-15}	10^{-18}	10^{-21}

5.4 Instrument calibration

Instrumental analytical techniques require **calibration** with **standards** of known analyte concentration. This establishes a relationship between the concentration of analyte in the sample and the signal measured by the instrument. It also ensures that results are traceable

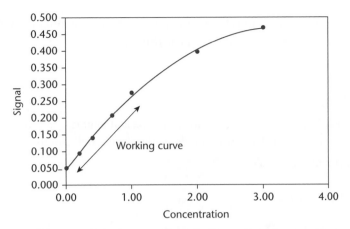

Figure 5.2 A calibration curve showing curvature at high concentration

and accurate. In order to do this a **calibration curve** must be obtained by plotting the **signal** as the dependent variable (on the *y*-axis), against **concentration** as the independent variable (on the *x*-axis). The term 'calibration curve', which has its basis in historical usage, is a somewhat misleading description because in modern methods of instrumental analysis the useful part of the calibration is generally regarded to be the linear region where, as we shall see later, the relationship between concentration and signal is well defined and reproducible. **Calibration function** is now the correct IUPAC term though calibration curve is still in common usage and is used throughout this book.

Often, the calibration curve will bend towards the concentration axis at higher concentrations, as shown in Figure 5.2.

As the calibration curves towards the *x*-axis at high concentration, the slope, and hence sensitivity, decreases. In addition, interpolation from the signal to the concentration axis becomes inherently imprecise, so it is preferable to use only the linear portion of the calibration known as the **working curve**. For example, in Figure 5.2 the signal does not increase as much between 2.0 and 3.0 concentration units as it does at lower concentrations, and if you were to extrapolate to 4.0 concentration units the signal would effectively be the same as it was at 3.0, clearly an unacceptable situation!

5.4.1 Preparing working standards

Preparation of **standard solutions** which are used for **calibration** is the most important step in the analytical method. Standards should be prepared that 'bracket' the concentrations likely to be found in the **prepared** samples, with standards of immediately lower and higher concentration. Samples should never be more concentrated than the strongest standard nor more dilute than the weakest.

Standards are usually prepared by **serial dilution** of a concentrated **stock solution** prepared from a **certified reference material** (see Chapter 6). The concentration of a diluted standard can be calculated using

$$c_{std} = \frac{V_{stock}}{V_{total}} \times c_{stock} \tag{5.1}$$

where c_{std} = concentration of the standard, V_{stock} = volume of stock solution, V_{total} = total volume, and c_{stock} = concentration of the stock solution.

Alternatively, if you wished to calculate the volume of stock solution needed to prepare a certain concentration then equation (5.1) can be rearranged to

$$V_{stock} = \frac{c_{std}}{c_{stock}} \times V_{total} \qquad (5.2)$$

Q: If you took 1 mL of a 1000 mg L^{-1} stock solution and diluted it to a total volume of 25 mL in a volumetric flask, what would be the concentration of the standard?

A: Use equation (5.1) to calculate c_{std}:

$$c_{std} = \frac{1.0}{25} \times 1000 = 40 \text{ mg L}^{-1}$$

Q: Alternatively, if you wished to prepare a 20 mg L^{-1} standard in a 100 mL volumetric flask then how would you calculate the volume of stock solution to pipette?

A: Use equation (5.2) to calculate V_{stock}:

$$V_{stock} = \frac{20}{1000} \times 100 = 2.0 \text{ mL}$$

ACTIVITY 5.1 Preparing working standards 1

You are required to prepare a series of standard solutions for the determination of nitrate in drinking water. The limit of detection of the technique (the lowest concentration that can be reliably measured—see Chapter 9) is 0.8 mg L^{-1} and the prepared samples typically range between 30 and 50 mg L^{-1}. The concentrated stock solution is 10 000 mg L^{-1}.

You have a choice of 100, 50, and 10 mL volumetric flasks and 1.0, 0.5, and 0.1 mL pipettes.

- How many and what concentration standards should you prepare?
- How much solution is required for analysis?
- What is the best way to calculate the dilutions given the equipment?

Record your results in a table like the one given below:

Standard	Concentration of stock/mg L^{-1}	Volume of stock/mL	Total volume/mL	Concentration of standard/mg L^{-1}
Std 1				
Std 2				
Std 3				
Std 4				
Std 5				
Std n				

When you have completed this task, compare your results with your tutor's and the feedback at the end of the chapter.

The example given in Activity 5.1 is relatively straightforward, and one that you would normally encounter in an undergraduate laboratory. However, when dealing with very small sample sizes it is often the case that you will perform serial dilution and manipulations solely using **micropipettes** to measure accurate volumes, or by **weighing**. To illustrate the point, take the example of morphine given in Section 5.2. The determination of morphine and its metabolites in blood is a common analytical procedure in clinical and forensic science for both qualitative identification and quantitative measurement. Clearly, you are unlikely to obtain a blood sample of more than a few mL and, given that the concentration of morphine could be at extremely low levels, dilution is undesirable. In this case then, preparation of standards and/or samples is best performed directly in clean 5 mL or 10 mL glass vials using 10–1000 µL micropipettes.

ACTIVITY 5.2 Preparing working standards 2

You are required to prepare a series of standard solutions for the determination of morphine by liquid **chromatography** mass spectrometry (LC-MS). The limit of detection of the technique is 10 ng mL^{-1} and the samples typically range between 100 and 500 ng mL^{-1}. The method requires that you prepare the standard solutions directly in the LC mobile phase. The concentrated stock solution is 10 000 ng mL^{-1}.

You have a choice of adjustable micropipettes in the ranges 10–100 µL, 10–250 µL, and 1.0–5.0 mL, and will be using 5 mL glass vials to prepare the standards. In this case you will need to pipette a known volume of chromatographic mobile phase into a vial and then a known volume of stock solution on top.

Record your results in a table like that given below:

Standard	Conc. of stock/ng mL^{-1}	Approx. conc. of standard/ ng mL^{-1}	Volume of stock/µL	Volume LC mobile phase/mL	Total volume/mL	Conc. of standard/ ng mL^{-1}
Std 1						
Std 2						
Std 3						
Std 4						
Std 5						
Std n						

5.4.2 Preparing a calibration curve

Best results are obtained when the standards are introduced to the instrument first in ascending order of concentration, and then in descending order. Modern instruments will normally have computer software for performing the calibration and calculating the concentration in the samples. However, it is important to understand the basic theory behind this process.

5.4.2.1 **Linear regression**

The most straightforward method of preparing a calibration curve is to plot concentration on the x-axis and signal on the y-axis and draw a line of best fit between the points. However, in order to reduce the subjective nature of drawing a line of best fit by eye, it is preferable to perform a linear regression which statistically models the linear relationship between the dependent variable (y data) and the independent variable (x data).

The aim is to plot a calibration for which there is a linear relationship between the x and y variables, so the curve has the form given by

$$y = bx + a \tag{5.3}$$

where b = gradient and a = intercept.

In addition to the gradient and intercept terms given in equation (5.3), computation of the linear regression will also yield the **product-moment correlation coefficient, r**, which is a measure of the linear dependence between the x and y variables. In the case of linear regression, another term, called the **coefficient of determination, R^2**, is often used (this is the term calculated by the 'trendline' function in Excel®), and in this case is simply the square of the correlation coefficient.

The values of r or R^2 can be calculated using a spreadsheet or by the instrument software. However, every analyst should be able to perform this task by hand so that they appreciate its limitations. For example, to calculate the product-moment correlation coefficient you must first calculate the residual for each data point, which is the difference between an individual x or y value and their means, that is, $x_i - \overline{x}$ and $y_i - \overline{y}$. After squaring, multiplying, and summing of these values, the information can then be substituted into an equation to calculate r. The process is most easily explained by an example.

ACTIVITY 5.3 Manually plotting a calibration curve, calculating the product-moment correlation coefficient, and fitting a linear regression

Plot the data given in Table 5.3 using graph paper but do not draw a line of best fit yet.

- The independent variable, concentration (which you control), must be plotted on the x-axis and the dependent variable, signal intensity, must be plotted on the y-axis.

- Label the axes with the variables' names and the units.

Table 5.3 Calibration data for morphine by LC-MS

Concentration/µg mL^{-1} x_i	Signal intensity/counts y_i
0.0	50
0.5	915
1.0	2150
2.0	3895
4.0	7750
8.0	15 900
16.0	31 505

→

→

Calculating the product-moment correlation coefficient

- Calculate the residuals and sums of the squares and record the data, as shown in Table 5.4.

Table 5.4 Data table for the residuals and sums of the squares

	x_i	y_i	$x_i - \bar{x}$	$(x_i - \bar{x})^2$	$y_i - \bar{y}$	$(y_i - \bar{y})^2$	$(x_i - \bar{x})(y_i - \bar{y})$
	0.0	50	−4.5	20.25	−8831...	77986561...	39740...
	0.5	915					
	1.0	2150					
	2.0	3895					
	4.0	7750					
	8.0	15 900					
	16.0	31 505					
Sums:							

- Calculate the product-moment correlation coefficient, r, using

$$r = \frac{\sum_i \{(x_i - \bar{x})(y_i - \bar{y})\}}{\left[\sum_i (x_i - \bar{x})^2 \sum_i (y_i - \bar{y})^2 \right]^{1/2}} \quad (5.4)$$

Fitting the line of regression of y on x

- Calculate the gradient, b, using

$$b = \frac{\sum_i \{(x_i - \bar{x})(y_i - \bar{y})\}}{\sum_i (x_i - \bar{x})^2} \quad (5.5)$$

- Calculate the intercept, a, using

$$a = \bar{y} - b\bar{x} \quad (5.6)$$

- Using the values for b and a plot the regression line of y on x on your original plot.

		Units
$r =$		
$b =$		
$a =$		

ACTIVITY 5.4 Plotting a calibration curve and fitting a regression trendline using Excel®

Curve 1

- Input the data shown in Table 5.5 into an Excel® spreadsheet.
- Plot the data in Excel® as a scatter chart.

Table 5.5 Calibration data for morphine by LC-MS

Concentration/μg mL^{-1} x_i	Signal intensity/counts y_i
0.0	50
0.5	915
1.0	2150
2.0	3895
4.0	7750
8.0	15 900
16.0	31 505

When you have created the chart, add a linear trendline, and display the R^2 value and equation of the line on the chart.

- Take the square root of R^2 to obtain r and compare the values for b, a, and r with the values calculated in Activity 5.3.

Curve 2

- Input the data shown in Table 5.6 into an Excel® spreadsheet.
- Plot the data in Excel® as a scatter chart.

Table 5.6 Calibration data for morphine by LC-MS

Concentration/μg mL^{-1} x_i	Signal/counts y_i
0	20
4	14 000
8	24 000
12	32 500
16	39 000
20	42 500

When you have created the chart, add a linear trendline and display the R^2 value and equation of the line on the chart.

- Does the regression line look like a good fit?
- Is there a linear relationship?
- Compare the R^2 value for curve 2 with that obtained for curve 1. Is it a good idea to only use the regression coefficient as an indicator of linearity?
- Is there a better relationship between the data points that you could plot manually?

It is evident from the activities that the correlation coefficient is a measure of the degree of linear relationship between two sets of data. However, the r value is open to misinterpretation. This is illustrated in Figure 5.3, which shows several plots obtained using a spreadsheet and their associated R^2 values (remember that this is the square of r, the product-moment correlation coefficient). As can be seen, the R^2 values alone would give the wrong impression of the underlying relationship, which may not necessarily be linear. Hence, it is essential to plot the data in order to check that linear least-squares statistics are appropriate in order to obtain accurate results (see the discussion in Chapters 7 and 9 regarding accuracy).

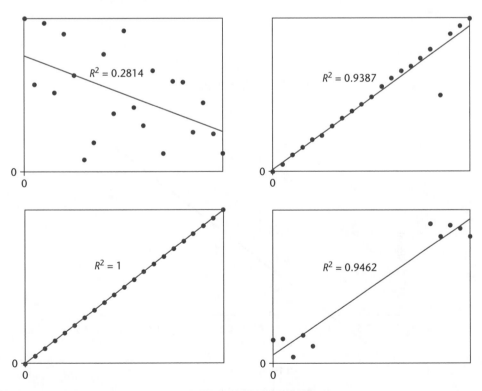

Figure 5.3 Data plots showing how the product-moment correlation coefficient is not necessarily a reliable measure of linearity

5.4.3 Standard additions

In Activity 5.2, the calibration standards were prepared directly in the LC mobile phase. If you have a supply of uncontaminated blood available a better way would be to prepare the standard solutions in blood; this is called **matrix matching** and is done to ensure that the standards and samples are similar in composition. This reduces interferences caused by the presence of the sample components other than the analyte, so called **matrix interferences**, during the sample preparation and/or analysis steps (matrix interferences are discussed in more detail in Section 5.6). An alternative to matrix matching is the method of **standard additions** which is useful when the composition of the sample is unknown or when an uncontaminated matrix diluent (e.g. a supply of blood which does not contain the analyte) is not available. The method can be used to correct for matrix interferences but not for spectral interferences or background changes (see Section 5.6.2.1).

In practice, at least three aliquots of the sample are taken. One is left untreated but the others are **spiked** with known additions of the analyte, preferably at about **0.5, 1, and 2 times** the concentration of the unknown. The volume of the addition should be negligible in comparison with the sample solution. This is to avoid dilution of the sample aliquots, which would alter the extent of the matrix interference and hence render the standard addition process invalid. The solutions are then analysed and a graph like that shown in Figure 5.4 is plotted. The line of best fit can be calculated in exactly the same way as in Activity 5.3 or by fitting a trendline using Excel®. The line is extrapolated back until it crosses the x-axis, the x intercept (i.e. x when $y = 0$) giving the negative of the concentration in the unspiked aliquot.

Hence, if the equation for the line of regression of y on x is known it can be used to calculate the concentration of analyte in the unknown by rearranging equation (5.3) and assuming that $y = 0$:

$$\text{Concentration in unknown} \quad x = -\frac{a}{b} \tag{5.7}$$

where a = intercept and b = gradient of the calibration.

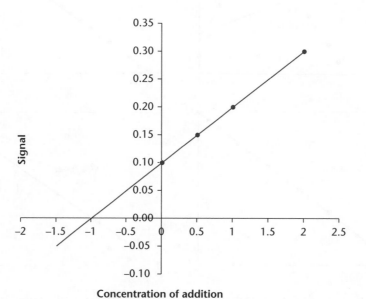

Figure 5.4 Method of standard additions

ACTIVITY 5.5 Creating a standard additions calibration curve

- Input the data shown in Table 5.7 into an Excel® spreadsheet.
- Plot the data as an Excel® scatter chart.
- Add a linear trendline and forecast it backwards until it intercepts the x-axis.
- From the equation of the trendline you can now extract the values for the slope and intercept and use them to calculate the concentration of morphine in the original sample using

$$\text{Concentration of morphine in sample} = -\frac{a}{b} \qquad (5.8)$$

Table 5.7 Standard additions data for morphine in blood by LC-MS

Added morphine/ng added per mL	Signal/counts
0	11 685
150	19 310
300	23 650
600	35 400

5.5 Quantitative analysis

Only when the standard solutions have been prepared and the instrument **calibrated** can the samples be analysed.

5.5.1 Using the calibration

The concentration of analyte in the procedural blank and sample solutions can be calculated either by **interpolation** from the calibration curve or by using the equation of the regression trendline as follows:

for normal calibration $x = \frac{y-a}{b}$ $\qquad (5.9)$

for standard additions $x = -\frac{a}{b}$ (because in this case $y = 0$) $\qquad (5.10)$

It is important to note that, in almost all cases, the procedural blank should be treated as if it were a sample, its concentration calculated, and this value subtracted from the concentration in each of the samples. The reason is that it gives an indication of the concentration of analyte in the reagents used to prepare the samples and general background contamination. If the concentration in the sample is very small compared to the procedural blank then negative values can result due to random errors in the measurement, that is, when a variable large number is subtracted from a slightly larger variable number (see Chapters 7 and 9).

ACTIVITY 5.6 Interpolating results from a calibration curve and calculating the concentration of analyte in the sample

You have used LC-MS to determine the concentration of morphine in blood and have calibrated the instrument using standard solutions prepared directly in the LC mobile phase. The blood samples themselves have been prepared by mixing 1.0 mL aliquots of blood with 2 mL of 10 mM ammonium carbonate buffer, pH 9 and centrifuged at 3000 rpm for 10 minutes. The supernatant is passed through a solid phase extraction (SPE) column which retains the analyte and separates it from the blood matrix. The analyte is eluted from the column in 3 mL of methanol, which is evaporated to dryness under a stream of nitrogen at 40°C, then reconstituted in 50 μL of the LC mobile phase, transferred to a 1 mL microcentrifuge tube, and centrifuged at 15000 rpm for 2 minutes. Aliquots (10 μL) are injected into the LC-MS and analysed. The results for three separate blood samples are shown in Table 5.8.

Table 5.8 Analysis data for morphine in blood samples

	Original volume of blood/mL	Instrument response/peak area counts
Procedural blank	0	125
Sample 1	1.0	7200
Sample 2	1.0	32600
Sample 3	1.0	1050

- Using the calibration data from Activity 5.3, determine the concentration of morphine in the sample, and hence the concentration in the blood sample. The a and b constants from the calibration curve are given below so you don't have to look them up from Activity 5.3.

  From Activity 5.3: $b = 1969.6 \ \mu g^{-1} \ mL$

  $a = 17.5$ counts

5.6 Sources of error

One of the biggest problems you will encounter when performing an analysis is how to deal with errors. We are not talking about errors in the statistical sense here, but rather gross errors or mistakes which can be due to poor sampling technique and/or preparation, contamination of the sample, loss of analyte, or instrumental errors.

5.6.1 Sampling and sample preparation

These topics are covered extensively in Chapters 2 and 3. However, it is worth reminding ourselves of the types of errors that can arise during the analysis stage if the sample has not been collected or prepared properly at the outset.

5.6.1.1 Sample homogeneity

The accuracy of an analysis depends critically on how representative the sample is of the material from which it is taken. When the analyte is spread evenly throughout the sample, such as when it is dissolved in a liquid, sampling is a relatively straightforward[2] process because any part of the liquid is probably representative of the whole and the sample is said to be **homogeneous**. However, when this is not the case, such as an unrefined ore, the sample is said to be **heterogeneous** and great care must be taken with sampling. Most instrumental analytical methods can typically be used on small samples (100 mg of solid or 10 mL of liquid), which increases the chances that the sample will not be representative. As described in the notes on sampling in Chapter 2, if either the concentration of the analyte in the sample does not represent that in the bulk material, or the concentration of the analyte in the solution at the time it is presented to the instrument has changed, the resultant error is likely to be greater than any other encountered during the actual analysis!

5.6.1.2 Contamination and losses

In most instrumental methods samples are presented for analysis as liquids, so solid samples must be dissolved or extracted. **Analytical** or **ultra-high purity** grade reagents must be used for dissolution or extraction to prevent contamination at trace and micro-trace levels.

Appropriate steps must be taken to prevent **analyte losses** due to volatilization or decomposition, for example, evaporation of volatile organic compounds from a mixture, volatilization of low-boiling-point metals from the sample during ashing in a muffle furnace, biological activity changing the concentration of oxygen in a sample of river water, and oxidation of vitamin C in a sample of orange juice. Analyte may also be lost by adsorption onto precipitates, such as the silica formed on **digestion** of silicate minerals using oxidizing acids.

Glassware, widely used for sample dissolution and preparation of standards, may give rise to further errors. Borosilicate glass, the type of glass used in standard laboratory grade glassware, has a surface that can act like an ion exchanger so, for trace level determinations, **contamination** and **losses** can occur through: (a) desorption or leaching of contaminants from the surface of the glass into solution; (b) adsorption of analyte from the solution onto the glass surface. To avoid such contamination all laboratory glassware should be cleaned using a method appropriate for the analysis. For trace metals analysis, glassware should be washed with a detergent and thoroughly rinsed, acid washed by soaking overnight in 10% V/V nitric acid (sp. gr. 1.41), then rinsed with deionized distilled water (DDW) and allowed to equilibrate in DDW overnight. This procedure ensures that any contaminants are first desorbed from the surface (by ion exchange with H^+ ions in the strong acid), then the subsequent equilibration with DDW passivates the surface to reduce the likelihood of subsequent re-adsorption from the standard or sample solution (even so, samples and standards should always be prepared in at least a 1% V/V strong acid to prevent re-adsorption of analyte). For organic analysis, glassware can be washed using chromic acid to oxidize any residual organic matter; however, contamination by specific organic compounds is less likely than by trace metals.

An alternative, which is becoming increasingly common, is to use disposable glass and plastic ware such as vials, centrifuge tubes, and pipette tips, which are now widely available, free from contamination. However, they should still be tested to make sure that they are **fit for purpose**.

For high sensitivity methods, further precautions may be necessary to avoid contamination, such as the use of clean rooms, laminar flow hoods, and special clothing to be donned before entering the laboratory. If ultra-low determinations are to be made then these precautions may be necessary at all times.

---

[2] Not necessarily always, such as in stratified natural systems, flowing streams or industrial processes.

Procedural blanks should be run for all analyses as a matter of course. A procedural blank has undergone the same preparation procedure as the sample, but without a sample being present, so it contains any analyte contamination arising from the reagents and equipment used. The concentration of analyte in the procedural blank can then be subtracted from the concentration in the sample (see Activity 5.6). Even if high purity reagents are used, the level of the analyte in the procedural blank may constitute the limiting factor in the analysis, and it may be necessary to purify reagents used for dissolution.

5.6.1.3 Chemical instability

Besides the obvious error of standards being wrongly made or sample preparation being bodged, it should not be forgotten that solutions and solids are often unstable. For example, standard solutions of 10 μg mL^{-1} and less often need to be prepared daily (largely due to degradation, adsorption onto the surface of the container or precipitation). Even 10 000 μg mL^{-1} stock solutions purchased from commercial suppliers will age and this is especially true when chemical changes can be expected in the analyte. Samples containing material of biological origin will obviously decompose unless preserved at low temperature or by desiccation. It is good practice to prepare and analyse all samples as swiftly as possible after collection; where this is not possible they should be stored in a freezer.

Often, but not always, the sample dissolution procedure stabilizes the analyte so the resultant solution can be stored at room temperature for several days or weeks. For example, a sample submitted for trace metals analysis can be digested in strong acid and stored at room temperature for several days; however, an extract of an organic compound is likely to degrade under the same conditions.

5.6.2 Instrumental errors

5.6.2.1 Non-spectroscopic interferences

Analytical instruments are no different to any other piece of electronic equipment: they take a while to warm up and do not take kindly to extremes of temperature or humidity. Normally, more complex instruments should be housed in an air conditioned laboratory with a stabilized power supply to ensure stable operation. However, at some point, the sample (usually a liquid) must be introduced into the instrument before a measurement can be made, and this step can critically affect the quality of the analysis. At the most basic level the measurement should only be commenced when the instrument has 'settled down' after a new sample has been introduced, which can take several seconds. Particular care must be taken when the concentration of the samples or standards changes markedly, especially if the new solution is more dilute because the measurement can be prone to **carry over** from the previous more concentrated sample.

Of most concern are the cases when the sample contains major concomitants in addition to the analyte, known as the **sample matrix**, and this can cause **non-spectroscopic** or **matrix interferences**. The type and severity of interference depends on the concentration and chemical composition of the matrix and the type of analysis being performed; it is impossible to cover every eventuality so several examples will serve to illustrate how they can be overcome. However, in general, non-spectroscopic interferences can generally be eliminated by:

- matrix matching the sample and standards;
- calibration using the method of standard additions;
- separating the sample matrix from the analyte;
- chemically modifying the sample *in situ*.

EXAMPLE 5.1

Matrix interferences during sample introduction in ICP-AES

In inductively coupled plasma atomic emission spectroscopy (ICP-AES), samples are introduced as liquids and a constant sample uptake rate is required to ensure comparability between samples and standards. The sample matrix can modify the viscosity, surface tension, and density of a sample solution compared to standard solutions, and this can affect the uptake rate; in the extreme case the sample interface can become blocked if the matrix concentration is very high. These factors influence the sample uptake rate, the aerosol particle size reaching the plasma, and the volatility of the analyte, which affects the stability of the analytical signal. An example of this is shown in Figure 5.5 where a check standard has been run between sample digests of an automobile catalytic converter for the determination of Pt by ICP-AES. As you can see, there is a serious degradation in the stability of the instrument after sample 24 has been introduced, and the instrument does not settle back down until sample 36 or so. This is probably due to a blockage in the sample introduction system.

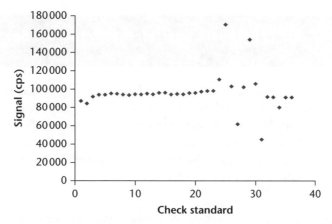

Figure 5.5 Atomic emission signal at 230.426 nm for a Pt check standard run during the determination of Pt in a catalytic converter digest by ICP-AES

EXAMPLE 5.2

Matrix interferences in the flame in FAAS

In flame atomic absorption spectrometry (FAAS), an example of this type of interference is that of phosphate on calcium. Here, the formation of a compound, probably calcium phosphate ($CaHPO_4$), which is less volatile (more refractory) than the halide or nitrate form of the analyte, hinders the formation of calcium atoms in the flame. The temperature of the flame (air–acetylene) is approximately 2200°C and is an important factor in this type of interference. Either EDTA can be added to the solution to competitively complex with Ca and form a more volatile compound, or another metal salt can be added to competitively complex with phosphate, thereby **releasing** the calcium from the interference. Another example of the formation of less volatile compounds is **occlusion** in a less volatile (e.g. refractory oxides of zirconium, uranium, etc.) or more volatile (ammonium chloride) compound. This can occur when a small amount of analyte becomes physically trapped in micro-particulates of the matrix and can depress or enhance the analyte signal respectively. A further example of an interference in FAAS is that caused by an easily ionized element (EIE) in the matrix. For example, imagine that you wish to determine sodium in a caesium matrix. Both elements are EIEs and will dissociate in the hot flame with the equilibria shown:

Analyte $Na \rightleftharpoons Na^+ + e^-$

Matrix (in excess) $Cs \rightleftharpoons Cs^+ + e^-$

The caesium (which is in great excess) will contribute by far the greater number of electrons so, by the law of mass action, the equilibrium for the dissociation of the analyte (Na) will be pushed to the left compared to the situation where no caesium is present (i.e. in the standard solution). In FAAS, absorption or emission of radiation by atoms, not ions, is measured so there will be an apparent increase in the concentration of sodium atoms in the sample compared to the standard for the same nominal concentration of analyte. The solution is to add an excess of an EIE to all standards, which is an example of a matrix matching procedure.

ACTIVITY 5.7 Phosphate interference on calcium determination by FAAS

You are a technician working for the Food Standards Agency and you are required to regularly test milk samples for calcium content using FAAS. The basic method requires that milk samples are diluted 50 times before direct analysis of the diluted sample by FAAS. Utilizing the data in Table 5.9, use a graphical method to illustrate the effect of phosphate on the absorbance signal for calcium with and without lanthanum present, interpret your graphs, and explain the effect of lanthanum.

Table 5.9 Absorbance data for a solution containing 20 mg L^{-1} dissolved calcium ions with varying amounts of phosphate with and without lanthanum present

Phosphate concentration/ mg L^{-1}	Absorbance signal for calcium	
	without La	with 100 mg L^{-1} La
0	0.330	0.329
10	0.295	0.330
15	0.278	0.315
20	0.245	0.324
25	0.226	0.323
30	0.199	0.319
35	0.180	0.310
40	0.175	0.315
50	0.171	0.318
60	0.170	0.311
80	0.172	0.310
100	0.171	0.304

5.6.2.2 Spectroscopic interferences

In spectroscopic measurements, interferences can arise from chemical species other than the analyte contributing to the analytical signal—for example, a matrix component emitting or absorbing radiation at the same spectral wavelength, or giving rise to an interfering ion at the same m/z, as the analyte of interest.

> **EXAMPLE 5.3**
>
> Spectroscopic interferences in mass spectrometry
> Morphine and hydromorphone are isomers having the same molar mass of 285.3377 g mol^{-1}. When analysed by mass spectrometry (MS) the two compounds produce the same molecular ion and exhibit very similar fragmentation patterns, as shown in Figure 5.6.
>
>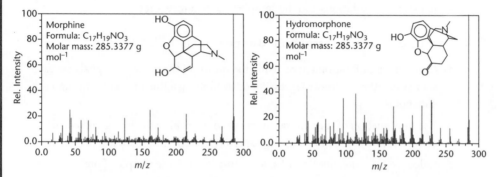
>
> Figure 5.6 Mass spectra of morphine and hydromorphone isomers
>
> Hence, spectroscopic interference between the two isomers is highly likely. The problem can be overcome by separating the two isomers prior to the MS analysis using liquid chromatography (LC), as shown in Figure 5.7.
>
>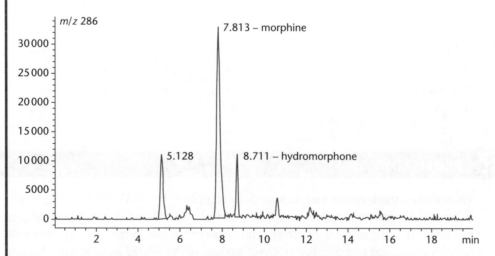
>
> Figure 5.7 Analysis of morphine and hydromorphone in blood by LC-MS
> (Source: © Agilent Technologies, Inc. 2005. Reproduced with Permission, Courtesy of Agilent Technologies, Inc.)

? PROBLEM 5.1

Analysis of arsenic ore

A sample of ore from a disused south-west Cornish mine was dried, ground, and sieved to less than 180 μm. A suitable quantity (0.9950 g) of this processed sample was then digested under reflux in a round bottomed flask on a heating mantle using 10 mL of *aqua regia* to extract the analyte (arsenic) contained within the ore. The resulting digest was filtered and the solution and washings collected in a 250.0 mL volumetric flask. The solution was made up to the mark and shaken. A 25.0 mL aliquot from this 250.0 mL volumetric solution was transferred to a 100.0 mL volumetric flask. This solution was then made up to the mark in order for the arsenic to be measured using inductively coupled plasma atomic emission spectroscopy (ICP-AES). A calibration was performed at 188.98 nm using a prepared series of standard arsenic solutions and the sample was then measured.

Complete the following tasks using the data provided in Table 5.10:

(a) Construct a calibration curve, using the means of the emission signals for each standard, either by linear regression or by plotting a line of best fit through the data points.

(b) Use the equation of the calibration curve to determine the blank-corrected concentration of arsenic in the measured (diluted) sample solution.

(c) Calculate the concentration of arsenic (mg kg^{-1}) in the sample of ore.

Table 5.10 Calibration and sample data for the determination of arsenic by ICP-AES

As concentration/mg L^{-1}	Emission signal from ICP-AES/counts				
	1	2	3	4	5
0	2	1	2	1	1
0.25	51	52	51	52	51
0.5	102	101	102	101	101
0.75	151	152	151	152	151
1.0	202	201	202	201	201
Procedural blank	67	68	66	67	67
Sample	122	121	122	121	121

? PROBLEM 5.2

Phosphate interference on calcium determination by FAAS

In Activity 5.7 you investigated phosphate interference on the determination of calcium by FAAS. The basic method for the determination of calcium in milk requires that milk samples are diluted 50 times before direct analysis of the diluted sample. Refer back to this activity and describe, giving reasons, whether you would need to adapt the basic method so you can perform an interference-free determination of calcium in a whole milk sample which has the typical composition shown in Table 5.11 (molar masses: $PO_4^{3-} = 94.97$ g mol^{-1}, P = 30.97 g mol^{-1}).

Table 5.11 Mineral composition of whole milk

Element	Content/mg per 100 mL of milk
Calcium	113
Copper	0.011
Iron	0.03
Magnesium	10
Manganese	0.003
Phosphorus	91
Potassium	143
Sodium	40
Zinc	0.40

Feedback on activities and problems

 FEEDBACK ON ACTIVITY 5.1

Preparing working standards 1
You are required to prepare a series of standard solutions for the determination of nitrate in drinking water. The limit of detection of the technique is 0.8 mg L^{-1} and the prepared samples typically range between 30 and 50 mg L^{-1}. The concentrated stock solution is 10 000 mg L^{-1}.

You have a choice of 100, 50, and 10 mL volumetric flasks and 1.0, 0.5, and 0.1 mL pipettes.

- *How many and what concentration standards should you prepare?*
  - You will need to prepare standards with concentrations from the limit of detection up to the top of the working curve, with most of them bracketing the likely concentration range of the samples. Five or six standard solutions are usually sufficient to prepare a linear calibration.

- *How much solution is required for analysis?*
  - This will depend on the type of instrument and the glassware available. A UV-vis spectrometer or autoanalyser will only require a few mL of solution and serial dilution using micropipettes into 10 mL volumetric flasks would work. However, the smallest volume glass pipette that should be used is 0.1 mL and this will limit the largest initial dilution factor to 100×. In the teaching laboratory when faced with an unfamiliar calibration it is probably best to decide on using the same volume volumetric flask for all standard solutions to simplify the dilutions, 50 or 100 mL the commonest sizes. For serial dilutions it will also probably be necessary to prepare an intermediate dilution of the stock solution to prepare the more dilute standards.

- *What is the best way to calculate the dilutions given the equipment?*
  - In order to simplify the calculations it is best to choose standard concentrations that can be easily prepared using the combination of pipettes and the chosen volumetric flask. There is no single correct way to do this, but an example is given

below where 100 mL volumetric flasks have been used throughout and different volume pipettes have been used to prepare standards which cover the appropriate concentration range.

Standard	Concentration of stock/mg L^{-1}	Volume of stock/mL	Total volume/ mL	Concentration of standard/mg L^{-1}
Std 1		0	100	0.0
Std 2	1000	0.10	100	1.0
Std 3	1000	0.50	100	5.0
Std 4	10 000	0.10	100	10
Std 5	10 000	0.50	100	50
Std 6	10 000	1.00	100	100

 FEEDBACK ON ACTIVITY 5.2

Preparing working standards 2

You are required to prepare a series of standard solutions for the determination of morphine by liquid chromatography mass spectrometry (LC-MS). The limit of detection of the technique is 10 ng mL^{-1} and the samples typically range between 100 and 500 ng mL^{-1}. The method requires that you prepare the standard solutions directly in the LC mobile phase. The concentrated stock solution is 10 000 ng mL^{-1}.

You have a choice of adjustable micropipettes in the ranges 10–100 µL, 10–250 µL, and 1.0–5.0 mL, and will be using 5 mL glass vials to prepare the standards. In this case you will need to pipette a known volume of chromatographic mobile phase into a vial and then a known volume of stock solution on top.

- How many and what concentration standards should you prepare?
  - You will need to prepare standards with concentrations from the limit of detection up to the top of the working curve, with most of them bracketing the likely concentration range of the samples. Five or six standard solutions are usually sufficient to prepare a linear calibration.

- How much solution is required for analysis?
  - In this case you are constrained by having to use 5 mL glass vials to prepare your standards. These are not volumetric glassware so all quantities must be measured using adjustable micropipettes. However, it is a good idea to choose solution volumes which are much smaller than the volume of mobile phase to be taken (to avoid dilution of the matrix) and that also keeps to a minimum the number of times that the pipettes have to be adjusted.

- What is the best way to calculate the dilutions given the equipment?
  - In order to simplify the calculations it is best to use the same volume of mobile phase for all the standards. LC-MS only requires at most a 1 mL sample, so 2.0 mL of mobile phase would be OK which could be pipetted with the 1.0–5.0 mL pipette. Next, choose standard concentrations that can be easily prepared using the minimum number

of combinations of pipette volumes. Normally, the minimum volume that can be pipetted accurately is 10 μL. Bearing this in mind, a typical example is given in the table below.

Standard	Conc. of stock/ng mL^{-1}	Approx. conc. of standard/ ng mL^{-1}	Volume of stock/μL	Volume LC mobile phase/mL	Total volume/mL	Conc. of standard/ ng mL^{-1}
Std 1	10 000	0	0	2.0	2.00	0
Std 2	10 000	50	10	2.0	2.01	50
Std 3	10 000	100	20	2.0	2.02	99
Std 4	10 000	250	50	2.0	2.05	244
Std 5	10 000	500	100	2.0	2.10	476
Std 6	10 000	1000	200	2.0	2.20	909

? FEEDBACK ON ACTIVITY 5.3

Manually plotting a calibration curve, calculating the product-moment correlation coefficient, and fitting a linear regression

- *Plot the data given in Table 5.3 using graph paper.*

You should have labelled a graph with concentration on the *x*-axis and signal intensity on the *y*-axis and plotted the data points but not the line of best fit, as shown below.

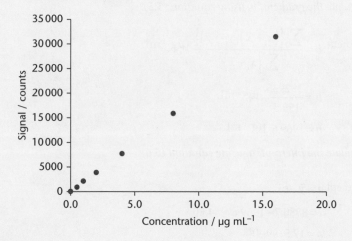

- *Calculate the residuals and sums of the squares and record the data, as shown in Table 5.4.*

A completed Table 5.4 should look like this:

Concentration/ μg mL^{-1} x_i	Signal/ counts y_i	$x_i - \bar{x}$	$(x_i - \bar{x})^2$	$y_i - \bar{y}$	$(y_i - \bar{y})^2$	$(x_i - \bar{x})(y_i - \bar{y})$
0.0	50	−4.5	20.25	−8831 …	77 986 561 …	39 740 …
0.5	915	−4	16.00	−7966 …	63 457 156 …	31 864 …
1.0	2150	−3.5	12.25	−6731 …	45 306 361 …	23 559 …
2.0	3895	−2.5	6.25	−4986 …	24 860 196 …	12 465 …
4.0	7750	−0.5	0.25	−1131 …	1 279 161 …	566 …
8.0	15 900	3.5	12.25	7019 …	49 266 361 …	24 567 …
16.0	31 505	11.5	132.25	22 624 …	511 858 304 …	260 176 …
Sums: 31.5	62 165	0	199.50	0	774 001 171 …	392 935

$$\text{Means} \quad \bar{x} = 4.50$$
$$\bar{y} = 8880.7$$

Note that the sums of the residuals in columns 3 and 5 sum to zero—this is to be expected because the data are spread equally either side of the mean.

- *Calculate the product-moment correlation coefficient, r, using equation (5.4):*

$$\text{Product-moment correlation coefficient } r = \frac{\sum_i \left\{ (x_i - \bar{x})(y_i - \bar{y}) \right\}}{\left[\sum_i (x_i - \bar{x})^2 \sum_i (y_i - \bar{y})^2 \right]^{1/2}}$$

$$r = \frac{392935}{\left[199.5 \times 774001171 \right]^{1/2}}$$

$$r = 0.99995$$

- *Calculate the gradient, b, using equation (5.5):*

$$\text{Gradient } b = \frac{\sum_i \left\{ (x_i - \bar{x})(y_i - \bar{y}) \right\}}{\sum_i (x_i - \bar{x})^2} \ \mu g^{-1} \ mL$$

$$b = \frac{392935}{199.5} \ \mu g^{-1} \ mL$$

$$b = 1969.6 \ \mu g^{-1} \ mL$$

- *Calculate the intercept, a, using equation (5.6):*

$$\text{Intercept } a = \bar{y} - b\bar{x}$$

$$a = 8880.7 - 1969.6 \times 4.50$$

$$a = 17.5 \ \text{counts}$$

- *Using the values for b and a plot the regression line of y on x on your original plot.*

In order to plot the regression line, which will be a straight line of best fit, you only need to know two points on the graph. One of the points is the intercept which has the coordinates (0, 17.5). The second point can be calculated from the equation of the line of best fit as follows:

Equation of the line $\quad y = bx + a$

Substituting for a and b $\quad y = 1969.6\ x + 17.5$

For $x = 16.0\ \mu g\ mL^{-1}$ $\quad y = 1969.6 \times 16.0 + 17.5$

$\qquad\qquad\qquad\qquad\quad y = 31\,531$ counts

The line of best fit should look like the graph below.

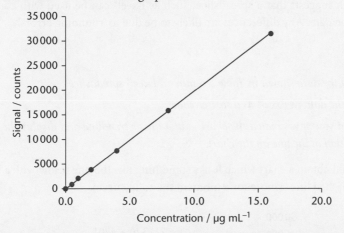

FEEDBACK ON ACTIVITY 5.4

Plotting a calibration curve and fitting a regression trendline using Excel®

Curve 1

- *Input the data shown in Table 5.5 into an Excel® spreadsheet.*
- *Plot the data in Excel® as a scatter chart.*
- *When you have created the chart, add a linear trendline and display the R^2 value and equation of the line on the chart.*

A description of how to plot an Excel® chart will not be given here because the instructions are likely to be out of date before publication, so I have assumed that you are able to do this part of the exercise and obtain a chart which looks like the one below.

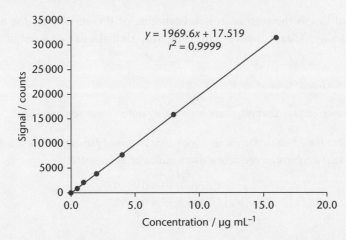

- *Take the square root of R^2 to obtain r and compare the values for b, a, and r with the values calculated in Activity 5.3.*

From the graph:

$$a = 17.5$$
$$b = 1969.6 \ \mu g^{-1} \ mL$$
$$r = \sqrt{R^2} = \sqrt{0.9999} = 0.99995$$

These values are in very close agreement with the values calculated manually in Activity 5.3, which suggests that a spreadsheet such as Excel® can be used with confidence to plot calibration data. Any differences are likely to be due to 'rounding errors'.

Curve 2

- *Input the data shown in Table 5.6 into an Excel® spreadsheet.*
- *Plot the data in Excel® as a scatter chart.*
- *When you have created the chart, add a linear trendline and display the R^2 value and equation of the line on the chart.*

You should obtain a chart which looks something like the one below with a linear trendline like the dashed line (ignore for a moment the solid curve).

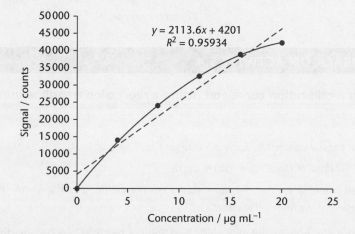

- *Does the regression line look like a good fit?*

The dashed line is the regression line (trendline or line of best fit, the terms are used interchangeably). Clearly there is something wrong and the line does not fit the data points very well.

- *Is there a linear relationship?*

In this case we can say that no, there is probably not a linear relationship.

- *Compare the R^2 value for curve 2 with that obtained for curve 1. Is it a good idea to only use the regression coefficient as an indicator of linearity?*

$$\text{Curve 1: } R^2 = 0.9999$$
$$\text{Curve 2: } R^2 = 0.9593$$

In both cases the R^2 value is close to 1, which suggests that a linear relationship exists for the data. However, we can see that this is clearly not the case for curve 2.

- *Is there a better relationship between the data points that you could plot manually?*

The solid curved line in the graph above is a much better fit to the data which suggests that the relationship is indeed not linear. This shows the importance of examining a graph to see if it actually shows what you think it does!

 FEEDBACK ON ACTIVITY 5.5

Creating a standard additions calibration curve

- *Input the data shown in Table 5.7 into an Excel® spreadsheet.*
- *Plot the data as an Excel® scatter chart.*
- *Add a linear trendline and forecast it backwards until it intercepts the x-axis.*

You should obtain a graph like the one below.

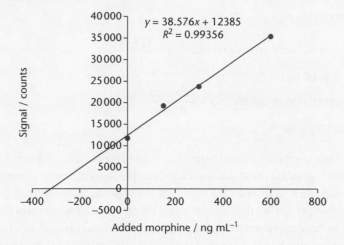

$y = 38.576x + 12385$
$R^2 = 0.99356$

- *From the equation of the trendline you can now extract the values for the slope and intercept and use them to calculate the concentration of morphine in the original sample using equation (5.8).*

From the graph:

$$a = 12\,385 \text{ counts}$$
$$b = 38.576 \text{ ng}^{-1} \text{ mL}$$
$$x = -\frac{a}{b} = -\frac{12385}{38.576} = -321$$

Concentration of morphine $= -321$ ng mL^{-1}

 FEEDBACK ON ACTIVITY 5.6

Interpolating results from a calibration curve and calculating the concentration of analyte in the sample

You have used LC-MS to determine the concentration of morphine in blood and have calibrated the instrument using standard solutions prepared directly in the LC mobile phase.

The blood samples themselves have been prepared by mixing 1.0 mL aliquots of blood with 2 mL of 10 mM ammonium carbonate buffer, pH 9 and centrifuged at 3000 rpm for 10 minutes. The supernatant is passed through a solid phase extraction (SPE) column which retains the analyte and separates it from the blood matrix. The analyte is then eluted from the column in 3 mL of methanol which is evaporated to dryness under a stream of nitrogen at 40°C; then reconstituted in 50 μL of the LC mobile phase, transferred to a 1 mL microcentrifuge tube, and centrifuged at 15 000 rpm for 2 minutes. Aliquots (10 μL) are injected into the LC-MS and analysed. The results for three separate blood samples are shown in Table 5.8.

• *Using the calibration data from Activity 5.3, determine the concentration of morphine in the sample, and hence the concentration in the blood sample.*

From Activity 5.3:

$b = 1969.6 \ \mu g^{-1} \, mL$

$a = 17.5 \ counts$

The first step is to calculate the concentration of the analyte in the procedural blank and each of the samples. For example,

Concentration: $x = \dfrac{y-a}{b}$

Concentration in procedural blank: $x = \dfrac{125 - 17.5}{1969.6}$

$x = 0.0546 \ \mu g \, mL^{-1}$

Concentration in sample 1: $x = \dfrac{7200 - 17.5}{1969.6}$

$x = 3.65 \ \mu g \, mL^{-1}$

The next step is to subtract the concentration of the procedural blank from the concentration of each sample to obtain the blank-subtracted sample concentration. A common error made by many students is to subtract the signal for the procedural blank from the signal for the samples before calculating the concentration. Under some circumstances this is acceptable; however, the concentration of the analyte in the procedural blank is an important piece of information because it tells the analyst the level of contamination in the sample preparation step. For example,

Blank subtracted conc. in sample 1 = 3.65 − 0.0546

= 3.60 $\mu g \, mL^{-1}$

The final step is to calculate the concentration of analyte in the original sample. In the method 1.0 mL aliquots of blood were taken and mixed with 2 mL of 10 mM buffer, centrifuged, passed through SPE, eluted with 3 mL of methanol, evaporated to dryness, reconstituted in 50 μL of mobile phase, transferred to a 1 mL centrifuge tube, centrifuged, and 10 μL aliquots were finally injected into the LC-MS and analysed—whew! The easiest way to do this is in a stepwise manner, working back from the final sample to the original. When you do this make sure that everything is in the same units, as follows:

Initial volume of sample: 1.0 mL

Intermediate dilutions: none

Final volume of sample: 50 μL = 0.050 mL

Conc. of morphine in final 0.05 mL volume = 3.60 $\mu g \, mL^{-1}$

Mass of morphine in final 0.05 mL volume $= 0.050$ mL $\times 3.60$ μg mL^{-1}

$$= 0.180 \text{ μg}$$

This mass of morphine came from the original 1.0 mL blood sample because there were no intermediate dilutions:

Mass of morphine in 1.0 mL blood sample $= 0.180$ μg

$$\therefore \text{Conc. of morphine in 1.0 mL blood sample} = \frac{0.180 \text{ μg}}{1.0 \text{ mL}}$$

$$= 0.18 \text{ μg mL}^{-1}$$

Results for all samples calculated in this way are summarized below.

	Original volume of blood/mL	Instrument response/ peak area counts	Concentration /μg mL^{-1}	Blank-sub-tracted sample concentration /μg mL^{-1}	Concentration in original blood sample /μg mL^{-1}
Procedural blank	0.0	125	0.0546		
Sample 1	1.0	7200	3.65	3.6	0.18
Sample 2	1.0	32600	16.5	16.5	0.825
Sample 3	1.0	1050	0.524	0.47	0.024

FEEDBACK ON ACTIVITY 5.7

Phosphate interference on calcium determination by FAAS

You should have plotted a graph that looks like the one below.

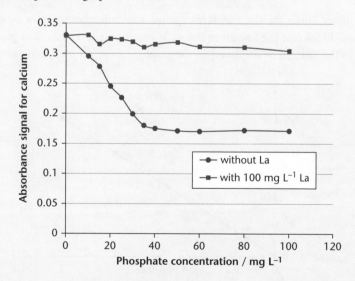

The graph shows that the absorbance signal for calcium decreases as the concentration of phosphate increases. However, in the presence of lanthanum the absorbance signal for calcium remains more or less constant between 0 and 100 mg L^{-1} phosphate.

In this case, calcium phosphate ($CaHPO_4$) is formed in the presence of high concentrations of phosphate. This is less volatile (more refractory) than the halide or nitrate form of calcium, so hinders the formation of calcium atoms in the flame. With lanthanum present the competitive formation of $LaPO_4$ releases Ca to form free atoms in the flame and the suppression of the Ca signal in the presence of phosphate is prevented. In practice, an excess of $LaCl_3$ would be added to both standards and samples.

  FEEDBACK ON PROBLEM 5.1

Analysis of arsenic ore

(a) *Construct a calibration curve, using the means of the emission signals for each standard, either by linear regression or by plotting a line of best fit through the data points.*

First calculate the means of the five replicate measurements of the standards, blank, and sample:

As concentration/mg L^{-1}	Emission signal from ICP-AES/counts					Mean
	1	2	3	4	5	
0	2	1	2	1	1	1
0.25	51	52	51	52	51	51
0.5	102	101	102	101	101	101
0.75	151	152	151	152	151	151
1	202	201	202	201	201	201
Procedural blank	67	68	66	67	67	67
Sample	122	121	122	121	121	121

Plot the mean values for the standards against concentration to get a calibration curve which should look like the one given below. If you used a spreadsheet to plot the data the equation of the line can be displayed on the graph.

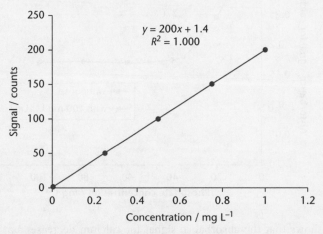

(b) *Use the equation of the calibration curve to determine the blank-corrected concentration of arsenic in the measured (diluted) sample solution.*

From the equation of the calibration:

$$b = 200 \text{ mg}^{-1} \text{ L}$$
$$a = 1.4 \text{ counts}$$

The first step is to calculate the concentration of arsenic in the procedural blank and the sample:

Concentration: $x = \dfrac{y-a}{b}$

Concentration in procedural blank: $x = \dfrac{67-1.4}{200}$

$x = 0.328 \text{ mg L}^{-1}$

Concentration in sample: $x = \dfrac{121-1.4}{200}$

$x = 0.598 \text{ mg L}^{-1}$

The next step is to subtract the concentration of the procedural blank from the concentration of the sample to obtain the blank-subtracted sample concentration:

Blank-subtracted conc. in sample $= 0.598 - 0.328$
$= 0.270 \text{ mg L}^{-1}$

(c) *Calculate the concentration of arsenic (mg kg^{-1}) in the sample of ore.*

The final step is to calculate the concentration of analyte in the original sample. In the method 0.9950 g of ore was digested in 10 mL of *aqua regia* and made up to 250.0 mL volume. A 25.0 mL aliquot of this was taken from this and diluted to 100.0 mL before analysis. To calculate the concentration in the original sample you must back-calculate taking into account any dilutions. The easiest way to do this is in a stepwise manner, working back from the final sample to the original. When you do this make sure that everything is in the same units, as follows:

Initial mass of sample: 0.9950 g
Initial volume of sample: 250.0 mL (after digestion)
Intermediate dilutions: 1 (25.0 mL taken from 250.0 mL)
Final volume of sample: 100.0 mL

Conc. of As in final 100 mL volume $= 0.270 \text{ mg L}^{-1}$

$= 0.270 \times 10^{-3} \text{ mg mL}^{-1}$

Mass of As in final 100 mL volume $= 100 \text{ mL} \times 0.270 \times 10^{-3} \text{ mg mL}^{-1}$

$= 0.0270 \text{ mg}$

This mass of As came from the 25.0 mL taken from the initial 250.0 mL volume during the intermediate dilution:

$\therefore$ Mass of As in 25.0 mL volume $= 0.0270 \text{ mg}$

$\therefore$ Mass of As in original 250.0 mL volume $= \dfrac{250.0 \text{ mL}}{25.0 \text{ mL}} \times 0.0270 \text{ mg}$

$= 0.270 \text{ mg}$

This mass of As came from the original 0.9950 g sample, so:

$$\text{Concentration of As in original sample} = \frac{0.270 \text{ mg}}{0.9950 \text{ g}}$$

Concentration of As in original sample = 0.271 mg g^{-1}

Concentration of As in original sample = 271 mg kg^{-1}

FEEDBACK ON PROBLEM 5.2

Phosphate interference on calcium determination by FAAS

The graph below shows that the absorbance signal for calcium decreases as the concentration of phosphate increases.

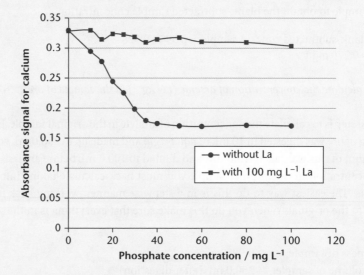

However, in the presence of lanthanum the absorbance signal for calcium remains more or less constant between 0 and 100 mg L^{-1} phosphate. In this case, calcium phosphate (CaHPO$_4$) is formed in the presence of high concentrations of phosphate. This is less volatile (more refractory) than the halide or nitrate form of calcium, so hinders the formation of calcium atoms in the flame. With lanthanum present the competitive formation of LaPO$_4$ releases Ca to form free atoms in the flame and the suppression of the Ca signal in the presence of phosphate is prevented. In practice, an excess of LaCl$_3$ would be added to both standards and samples.

In order to decide whether the milk samples would be prone to interference it is necessary to calculate the concentration of PO$_4^{3-}$ in the milk as follows:

Concentration of P = 91 mg/100 mL

$\qquad\qquad = 91 \times 10 \text{ mg L}^{-1}$

$\qquad\qquad = 910 \text{ mg L}^{-1}$

Molar mass of P = 30.97 g mol^{-1}

Concentration of P = 910/30.97 mmol L^{-1}

$$= 29.38 \text{ mmol L}^{-1}$$

Molar mass of PO$_4^{3-}$ = 94.97 g mol^{-1}

Concentration of PO$_4^{3-}$ = 29.38 × 94.97 mg L^{-1}

$$= 2782 \text{ mg L}^{-1}$$

After 50× dilution,

Concentration of PO$_4^{3-}$ = 2782/50 mg L^{-1}

$$= 56 \text{ mg L}^{-1}$$

At this concentration of PO$_4^{3-}$ the signal for Ca will be suppressed by approximately 50%. In order to overcome this, lanthanum at a concentration of 100 mg L^{-1} should be added to act as a releasing agent.

Reference materials and standards

Learning outcomes

Once you have read through this chapter, worked through the activities and problems, and reviewed the feedback, you should be able to:

1. distinguish between the different types of reference materials;

2. understand the importance of traceability in analytical chemistry;

3. select an appropriate reference material for use in a variety of scenarios in analytical chemistry.

. . . .

6.1 **What are reference materials?**

Before going on to describe reference materials in detail, it should be noted that this chapter is complementary to Chapters 7 and 9, and its relevance will be more apparent after reading these.

The definition and uses of reference materials are set out in a series of publications[1] by the International Organization for Standardization (ISO):

- *ISO Guide 30:1992 Terms and definitions used in connection with reference materials;*

- *ISO Guide 30:1992/Amd 1:2008 Revision of definitions for reference material and certified reference material;*

- *ISO Guide 32:1997 Calibration in analytical chemistry and use of certified reference materials;*

- *ISO Guide 33:2000 Uses of certified reference materials.*

---

[1] ISO. 2012. *Standards Catalog* Available at: http://www.iso.org/iso/home/store/catalogue_tc/catalogue_tc_browse.htm?commid=55002&published=on

Reference materials are defined in *ISO Guide 30:1992/Amd 1:2008* as follows:

- Reference material (RM)
  . . . material, sufficiently homogeneous and stable with respect to one or more specified properties[2], which has been established to be fit for its intended use in a measurement process.[3]

- Certified reference material (CRM)
  . . . reference material characterized by a metrologically valid procedure for one or more specified properties, accompanied by a certificate that provides the value[4] of the specified property, its associated uncertainty, and a statement of metrological traceability.[5]

In common parlance, a reference material is any substance that has been characterized for a particular chemical or physical property. This property can be quantitative (e.g. the concentration of lead) or qualitative (e.g. a DNA sequence). Reference materials can be used for instrument calibration or method validation, but a single RM cannot be used for both of these within the same procedure.

6.2 Traceability

Inherent in the use of a reference material is the concept of **traceability**, such that an analytical measurement is ultimately traceable to SI units. You are probably familiar with the SI unit of the metre. The original traceable standard for this was a prototype metre bar made from 90% Pt and 10% Ir—the distance between marks on the bar at 0°C was exactly 1 m. In 1960 this was replaced with a new definition of 1 650 763.73 times the wavelength of radiation emitted by the [86]Kr isotope, then again in 1983 as the length of the path travelled by light in vacuum during a time interval of 1/299 792 458 s. In the case of chemical analysis, measurement traceability is to two other SI units:

- **amount of substance**, that is, the mole (mol);
- **mass**, that is, the kilogram (kg).

Q: How could you ensure traceability of a standard solution, prepared from a pure chemical in the laboratory?

A: In order to prepare a standard solution you would first need to measure accurately a known mass of the pure chemical using an analytical balance. Assuming that the laboratory was following quality control procedures, the balance will have been calibrated using a set of standard masses. These masses are **reference materials** which can be further calibrated against **certified reference material** masses, and ultimately to the international prototype kilogram, which is conserved and used at the **International Bureau of Weights and Measures (BIPM)** in Sèvres, France.

---

[2] Properties can be quantitative or qualitative, for example, identity of substances or species.

[3] Uses may include the calibration of a measurement system, assessment of a measurement procedure, assigning values to other materials, and quality control; a single RM cannot be used for both calibration and validation of results in the same measurement procedure.

[4] The concept of value includes qualitative attributes such as identity or sequence. Uncertainties for such attributes may be expressed as probabilities.

[5] Metrologically valid procedures for the production and certification of reference materials are given in, among others, ISO Guides 34 and 35.

6.3 Uses of reference materials

Reference material (RM) is a generic term which applies to two categories of materials which can be used for different purposes:

1. **Certified reference material (CRM)**—these can be either a matrix material or a single substance:
   - **Single substance:** a single material certified for either its chemical purity (e.g. a pesticide) or for a specific property (e.g. material of known melting point). These types of material are useful for instrument calibration but are not generally used for method validation, for example.
   - **Matrix material:** a complex material with one or more of the analytes in the matrix known with accuracy (e.g. an oil sample containing a known amount of sulfur). Providing that a CRM with a suitable matrix can be obtained, these types of material are suitable for method validation.

Both of the above have undergone a stringent process of characterization, usually at multiple laboratories, and are accompanied by a certificate of analysis which lists the value of the property which has been measured plus an estimate of the uncertainty. An example of a single substance CRM, an NO_3^- chromatography standard, is shown in Figure 6.1; notice that the certificate gives information about the concentration, uncertainty, and traceability. The certificate for a matrix CRM is shown in Figure 6.2; here the concentrations of eleven different compounds in a polypropylene matrix have been determined and certified.

2. **Reference material (RM)**—this pretty much covers everything else prepared using CRMs to make a secondary set of reference materials which include:
   - in-house reference materials—materials prepared in the laboratory and characterized using a certified reference material for calibration;
   - quality control materials—materials with a known concentration of analyte used routinely to check the accuracy of the method;
   - spiked samples—a sample to which a known quantity of a single substance reference material has been added.

Which of these materials are appropriate for use at any one time will depend on the requirements of the analyst. Some common uses for reference materials are:

- method validation (Chapter 9);
- instrument calibration (Chapter 5);
- verification of instrument, laboratory, and analyst performance (Chapter 9);
- uncertainty estimation;
- internal quality control (Chapter 9).

So, we can see that both types of reference material are essential to the analytical process. Whereas a standard solution is necessary for calibrating the response of an instrument, a matrix reference material is equally necessary to assess the suitability of an analytical method. Both CRMs and other reference materials should be used on a regular basis as part of a quality assurance system in order to ensure that the analytical method is under statistical control and is producing results that are **fit for purpose**.

C E R T I F I C A T E   O F   A N A L Y S I S

Ion Chromatography CRM

Nitrate (NO₃⁻) – 1000 µg/mL

Matrix: H_2O

Product #: VHG-INO3-500   Lot #: 101498-10

Ion	Certified Concentration & Uncertainty
NO₃⁻	1000 ± 5 µg/mL

Intended Use: This solution is intended for use as a certified reference material or calibration standard, for ion chromatography (IC) or techniques using other modes of aqueous ion detection.

Certification & Traceability: VHG CRMs are manufactured and certified under a quality management system that is accredited to ISO 9001, ISO Guide 34 and ISO/IEC 17025. This CRM was prepared to a nominal concentration of 1000µg/mL by gravimetric methods using 99.99% pure sodium nitrate (NaNO₃) dissolved and diluted with filtered (0.22µm), 18 M-ohm deionized water. The balances used in the preparation of VHG CRMs are calibrated regularly with traceability to NIST. All volumetric dilutions are performed in Class A calibrated glassware. The certified concentration was determined by VHG Labs based upon gravimetric procedures. Secondary verification of the certified concentration was performed by VHG Labs using ion chromatography (IC) or inductively coupled plasma optical emission spectroscopy (ICP-OES), which was calibrated and/or referenced against NIST SRM 3185. The uncertainty associated with the certified concentration represents the expanded uncertainty at the 95% confidence level using a coverage factor of k=2.

Instructions for Use: We recommend that the solution be thoroughly mixed by repeated shaking or swirling of the bottle immediately prior to use. To achieve the highest accuracy the analyst should: (1) use only pre-cleaned containers and transferware, (2) not pipette directly from the CRM's original container, (3) use a minimum sub-sample size of 500µL, (4) make dilutions using calibrated balances or certified volumetric class A flasks and pipettes, (5) dilute with the same matrix as the original CRM, and (6) never pour used product back into the original container. The solution should be kept tightly capped and stored under normal laboratory conditions. Do not freeze, heat, or expose to direct sunlight. Minimize exposure to moisture or high humidity.

Period of Validity: VHG ensures the accuracy of this solution for **12 Months** from the Certification date shown below, provided the instructions for use are followed. During the period of validity, the purchaser will be notified if this product is recalled due to any significant changes in the stability of the solution.

VHG Labs, Inc.

Susan Evans Norris, Certifying Officer

See Exp. on Container
Certification Date

VHG Labs, Inc. waives all responsibility for any damages resulting from the usage and/or implementation of the products/data described herein.

Figure 6.1 Certificate for an NO₃⁻ chromatography standard, an example of a single substance CRM

6.3.1 Availability of reference materials

Reference materials are available from a wide range of sources. However, higher order certified reference materials are only available from an accredited producer (e.g. accredited to ISO Guide 34)[6] such as a **National Metrology[7] Institute (NMI)**. Such CRMs are accompanied by detailed certificates of analysis, uncertainties, and metrological traceability of the certificated properties, instructions for storage, handling and use. A list of NMIs can be obtained from the BIPM on their website (www.bipm.org). Some of the better-known producers of CRMs are given below, though this list is by no means exhaustive:

LGC   National metrology institute of the UK (http://www.lgc.co.uk/), the largest supplier of reference materials in Europe, which also produces a wide range of RMs and CRMs in collaboration with BAM and IRMM under the European Reference Material (ERM) brand.

---

[6] ISO Guide 34:2009. General requirements for the competence of reference material producers.
[7] Metrology: field of knowledge concerned with measurement.

CERTIFICATE OF ANALYSIS

ERM®- EC591

POLYPROPYLENE		
	Mass Fraction	
	Certified value [2] [g/kg]	Uncertainty [3] [g/kg]
Br	2.08	0.07
2,4,4'-TriBDE (BDE-28)[1]	0.0025	0.0004
2,2',4,4'-TetraBDE (BDE-47)[1]	0.245	0.023
2,2',3,4,4'-PentaBDE (BDE-99)[1]	0.32	0.04
2,2',4,4',6-PentaBDE (BDE-100)[1]	0.066	0.007
2,2',4,4',5,5'-HexaBDE (BDE-153)[1]	0.044	0.006
2,2',4,4',5,6'-HexaBDE (BDE-154)[1]	0.026	0.004
2,2',3,4,4',5,6'-HeptaBDE (BDE-183)[1]	0.087	0.008
2,2',3,3',4,4',6,6'-OctaBDE + 2,2',3, 4,4',5,6,6'-OctaBDE (BDE-197+204)[1]	0.052	0.009
DecaBDE (BDE-209)[1]	0.78	0.09
DecaBB (BB-209)[1]	0.74	0.08

1) Brominated diphenylethers (BDE) or brominated biphenyls (BB) as obtained by quantification by GC-MS. The numbering is according to the system by Ballschmiter et al. (J. high resolut. chrom. 15:260-270).

2) Unweighted mean value of the means of accepted sets of data, each set being obtained in a different laboratory and/or with a different method of determination. The certified value and its uncertainty are traceable to the International System of Units (SI).

3) The certified uncertainty is the expanded uncertainty estimated in accordance with the Guide to the Expression of Uncertainty in Measurement (GUM) with a coverage factor k = 2, corresponding to a level of confidence of about 95 %.

This certificate is valid for one year after purchase.

Sales date:

The minimum amount of sample to be used is 30 mg for Br and the polybrominated flame retardants and 200 mg for Sb.

NOTE

European Reference Material ERM®-EC591 was produced and certified under the responsibility of the IRMM according to the principles laid down in the technical guidelines of the European Reference Materials® co-operation agreement between BAM-IRMM-LGC. Information on these guidelines is available on the internet (http://www.erm-crm.org).

Accepted as an ERM®, Geel, December 2008
Latest revision: May 2009

Signed: _____

Prof. Dr. Hendrik Emons
European Commission
Joint Research Centre
Institute for Reference Materials and Measurements
Retieseweg 111
B-2440 Geel, Belgium

All following pages are an integral part of the certificate.

Page 1 of 4

Figure 6.2 Certificate for a mixture of fire retardants in polypropylene, an example of a matrix CRM

BAM   Bundesanstalt für Materialforschung (Federal Institute for Materials Research and Testing, Germany, http://www.bam.de/)

IRMM   Institute for Reference Materials and Measurements. Joint research centre of the European Commission (https://irmm.jrc.ec.europa.eu/)

NIM   National Institute of Metrology, China (http://en.nim.ac.cn/)

NIST   National Institute of Standards and Technology, USA (http://www.nist.gov)

NMIA   National Measurement Institute of Australia (http://www.measurement.gov.au/)

NMIJ   National Metrology Institute of Japan (http://www.nmij.jp/english/)

6.3.2 Selection of reference materials

Selection of a reference material for a particular purpose will depend on the nature of the activity, so it is the job of the analyst to assess this. The following factors should be considered when selecting a reference material:

- How closely does the matrix match the sample (sample preparation and potential interferences will depend on this)?
- Does it contain the analyte to be determined?
- Is the analyte at a similar concentration compared to the sample (if it is too far outside the measurement range this may be inappropriate)?
- Is the measurement uncertainty[8,9] appropriate (i.e. usually not too large)?
- Are the certification procedures used by the producer appropriate (i.e. what class of RM is it)?
- Is there appropriate documentation supplied with the material (e.g. certificate, report, etc.)?

ACTIVITY 6.1 Choosing an appropriate reference material

ERM-AD425 Soybean DNA	ERM-CA615 Trace metals in groundwater	BCR-326 Zinc (unalloyed)
NIST SRM-3103a Arsenic solution	CIL-BDE-100-CS 2,2',4,4',6-Pentabromodiphenyl ether (BDE-100) in nonane	ERM-BF425d Soybean powder
NIST SRM-682 Zinc (metal)	ERM-CA616 Simulated groundwater	ERM-EC591 Fire retardants in polypropylene

→

[8] UK Accreditation Service. 2013. *Measurement Uncertainty*. Available at: https://www.ukas.com/services/technical-services/technical-articles/1250-2/

[9] Measurement uncertainty is 'a parameter, associated with the result of a measurement (e.g. a calibration or test) that defines the range of the values that could reasonably be attributed to the measured quantity. When uncertainty is evaluated and reported in a specified way it indicates the level of confidence that the value actually lies within the range defined by the uncertainty interval' (UK Accreditation Service, 2013).

→

Refer to the certificates of the reference material(s) in the grid (see the online resources for links to these certificates) and then describe how you would use them in the following situations:

(a) calibration of an instrument for the determination of Zn in a sample of zinc ore;

(b) calibration of a method for the determination of the amount of genetically modified soya contained in a batch of dried soybean bought from the supermarket;

(c) validation of a method for the determination of BDE-100 fire retardant in polypropylene carpet;

(d) quality control of a method for the determination of As in groundwater in the concentration range $0.01-0.1$ $\mu g\ L^{-1}$.

 PROBLEM 6.1

Determination of arsenic in beef

You have just started a new job as quality control manager at a food cannery. As one of your first tasks, you have been asked to evaluate a new method for the determination of 'total arsenic' in beef products. Initial findings indicate that you will be measuring concentrations of 'total arsenic' in beef in the range $0.01-0.1$ $\mu g\ g^{-1}$. The published analytical method you propose to use consists of three steps:

1. an acid digestion step: microwave bomb digestion of the sample beef with concentrated acid;

2. an oxidation/reduction step: to convert the arsenic to a form that can be reacted to give a volatile hydride;

3. an analysis step: analysis of sample solutions using sodium borohydride ($NaBH_4$) hydride generation atomic fluorescence spectrometry.

The reference materials which you have available are given in Table 6.1. In addition, you have a supply of freeze-dried, uncontaminated beef supplied by the Food Standards Agency (FSA). Explain how you might use these reference materials in the following situations:

(a) Calibration of the instrument using arsenic solutions of known concentration made up in a similar matrix to the samples;

(b) Production of analytical data to show that the method you are using is performing satisfactorily;

(c) Preparation of an in-house reference material to check that you are getting accurate results;

(d) When the method was previously used to measure 'total arsenic' in tuna, several low results were reported. The explanation given for the poor results was that some of the arsenic was present as arsenobetaine, which does not react to form a volatile hydride. You are required to check to see whether the same thing is likely to happen again.

Table 6.1 Reference materials for use in the analytical method

Material	Matrix	Total arsenic concentration /µg g^{-1}	Uncertainty** /µg g^{-1}
Certified Reference Material BCR185R	Bovine liver	0.0330	± 0.0029*
Certified Reference Material NIST8414	Bovine muscle	0.009	± 0.003*
Certified Reference Material NIST3103a	Inorganic arsenic standard solution in 10% nitric acid	9941	± 55*
Certified Reference Material BCR626	Arsenobetaine standard solution in water	1031	± 6*

* Half-width of the 95% confidence interval of the mean.
** Uncertainty in the certified value (ISO definition).

Feedback on activities and problems

 FEEDBACK ON ACTIVITY 6.1

Choosing an appropriate reference material

(a) *Calibration of an instrument for the determination of Zn in a sample of zinc ore*

You would need to prepare a solution of pure Zn metal to use as a stock solution for your calibration standards. This can be done by weighing an accurate mass of the pure zinc metal CRM 'NIST SRM-682 Zinc', and dissolving it in acid to an accurately known volume or mass. It would not be appropriate to use 'BCR-326 Zinc' because this is a matrix CRM containing certified concentrations of Cd, Cu, Fe, and Pb, more suited to validation, for example, of a method for the determination of these elements in unalloyed zinc.

(b) *Calibration of a method for the determination of the amount of genetically modified soya contained in a batch of dried soybean bought from the supermarket.*

Both 'ERM-AD425 Soybean DNA' and 'ERM-BF425d Soybean powder' could be used to calibrate the method because both materials contain a certified concentration of genetically modified '356043 soya'. The difference between them is that ERM-AD425 is the pure DNA plasmid fragment whereas ERM-BF425d is a mixture of modified and unmodified soya in the form of dried soybean powder. If method validation was required then ERM-BF425d could also be used as a matrix CRM for this purpose.

(c) *Validation of a method for the determination of BDE-100 fire retardant in polypropylene carpet*

The matrix CRM 'ERM-EC591 Fire retardants in polypropylene' contains a certified concentration of BDE-100, so could be used for method validation. Calibration of the instrument could be performed using CIL-BDE-100-CS because this is a solution of pure BDE-100 in nonane and can be diluted to prepare a series of calibration solutions.

(d) *Quality control of a method for the determination of As in groundwater in the concentration range 0.01–0.1 µg L^{-1}.*

ERM-CA615 Trace metals in groundwater would seem to be most suitable for this, however, the CRM is certified for an As concentration of 9.9 µg L^{-1}, which is much higher than the measurement range of the method. A better approach may be to spike the simulated groundwater CRM ERM-CA616 with a known concentration of the NIST SRM-3103a Arsenic solution to prepare a secondary reference material which contains arsenic in the correct concentration range. The NIST SRM-3103a can also be used to calibrate the instrument because it is a solution of pure arsenic intended for this use.

 FEEDBACK ON PROBLEM 6.1

Determination of arsenic in beef

(a) *Calibration of the instrument using arsenic solutions of known concentration made up in a similar matrix to the samples*

You can use the NIST3103a CRM to prepare standard solutions spiked into digested samples of the uncontaminated beef supplied by the FSA. This ensures that the calibration solutions are matrix-matched with the samples. Alternatively, you could analyse the samples by the method of standard additions after spiking with NIST3103a.

(b) *Production of analytical data to show that the method you are using is performing satisfactorily*

You can show that you are getting the right answers from your analyses by analysis of a suitable certified reference material which contains a known amount of total arsenic. You have two certified reference materials available, BCR185R and NIST8414. The former contains As at the correct concentration for the analysis (in the range 0.01–0.1 µg g^{-1}) but the liver matrix is not exactly the same as the sample matrix; the latter contains As at a slightly lower concentration than the samples to be tested but is likely to be more similar in composition. In practice, you would probably need to use both materials.

(c) *Preparation of an in-house reference material to check that you are getting accurate results*

You can spike uncontaminated beef using the NIST3103a CRM and analyse this as if it were a sample. In this way you can check that the method is performing satisfactorily without using the expensive matrix CRMs.

(d) *When the method was previously used to measure 'total arsenic' in tuna, several low results were reported. The explanation given for the poor results was that some of the arsenic was present as arsenobetaine, which does not react to form a volatile hydride. You are required to check to see whether the same thing is likely to happen again.*

Arsenic can be present in a sample in a number of different chemical forms, or species, typically As(III) (arsenite or AsO_3^{3-}), As(V) (arsenate or AsO_4^{3-}), dimethylarsenic acid (DMAA or Me_2AsO_2H), and methylarsonic acid (MAA or $MeAsO_3H_2$). In addition, arsenic frequently occurs in fish in the form of arsenobetaine (trimethylarsenioacetate or $Me_3As^+CH_2COO^-$).

The reaction scheme for hydride generation using $NaBH_4$ is:

$$3NaBH_4 + 4H_3AsO_3 \rightarrow 4AsH_{3(g)} + H_3BO_3 + 3NaOH$$

So, in order to form the volatile hydride arsine (AsH_3) quantitatively it is necessary to ensure that all of the arsenic in the samples and standards is in the As(III) state. Unfortunately, arsenobetaine is particularly resistant to chemical breakdown so it may persist even after sample digestion and oxidation/reduction. In order to check whether this is happening, the BCR626 arsenobetaine standard can be used to check where problems might arise by spiking uncontaminated beef samples at different stages in the method.

Basic statistics

Learning outcomes

Once you have read through this chapter, worked through the activities and problems, and reviewed the feedback, you should be able to:

1. use a range of methods to display analytical data;

2. understand a range of basic statistical concepts;

3. explain how to use a *t*-test to compare a mean with a true value and comparison of two means;

4. use appropriate statistical tests and graphical methods to evaluate and interpret analytical data in a variety of contexts.

7.1 Introduction

There are few things more important to our wellbeing than water. In the developed world we tend to take it for granted that water which comes out of the tap is fit to drink. However, the **quality** of this water depends on a chain of factors: how plentiful is the supply; the extent of pollution in rivers, lakes, and groundwater; the effectiveness of drinking water treatment procedures; and the infrastructure used to deliver the final product to our homes. At all of these stages the water must be checked to make sure its quality falls within statutory limits.

In the lakes and rivers of western Europe, one of the problems commonly encountered is **eutrophication**, the enrichment of a water body with nutrients such as nitrate (NO_3^-) and orthophosphate (PO_4^{3-}). Eutrophication can lead to algal blooms in estuaries and coastal waters, which have become increasingly common in the UK and around the world. This type of increased biological activity is harmful because it causes the dissolved oxygen to be used up with the consequence that fish cannot survive and, in extreme cases, waters become anaerobic and 'stagnant'.

In drinking water the level of nitrate (NO_3^-) must be controlled because it is toxic at high concentration and can lead to a condition called methemoglobinemia (blue-baby syndrome) in bottle-fed infants. The drinking water regulations for the UK are based on European standards and set a **maximum admissible concentration** (MAC) of nitrate in drinking water.

ACTIVITY 7.1 Finding regulatory information on drinking water

Use the World Wide Web to locate the **Drinking Water Inspectorate** website and access the consumer advice leaflet *Drinking water standards/regulations*. Write down the MAC for nitrate:

MAC for nitrate (NO_3^-):   _____ mg L^{-1}

When you have completed this activity, study the feedback given at the end of the chapter.

7.2 **Visualizing data**

Analytical scientists have to make decisions based on the results of the measurements they make. For example, how can you tell whether the concentration of nitrate in drinking water in a particular sample is above the MAC? Sometimes the answer is clear cut, but often it is not so obvious, so we need a way of comparing measurements to support any decisions we make. In order to do this we must use **statistics**, but first it is often useful to study the data visually using some of the following methods:

7.2.1 **Blob plots**

For small datasets, up to 10 data points, blob plots can be used. These are made by plotting each result, as a blob, on a horizontal scale.

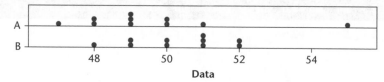

7.2.2 **Histograms**

For larger datasets histograms are more useful. The vertical axis gives the number of results that fall within a small defined range on the horizontal measurement scale.

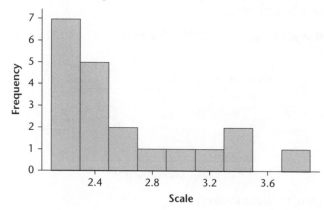

7.2.3 Box and whisker plots

When datasets become larger it is easier to summarize them in the form of a box and whisker plot (Figure 7.1).

- The **median** (Q2) is the mid-point of the data, so 50% lie above and below this line.
- The **interquartile range** (IQR) is the middle 50% of the data, and is bounded by the **lower quartile** Q1 (25% of the data are less than or equal to this) and the **upper quartile** Q3 (75% of the data are less than or equal to this), hence IQR = Q3 − Q1.
- The upper whisker extends to the maximum data point within 1.5 box heights from the top of the box.
- The lower whisker extends to the minimum data point within 1.5 box heights from the bottom of the box.
- An **outlier** (*) is a datum beyond the upper or lower whisker.

Box and whisker plots are helpful to get an idea of the spread and skewness of a distribution and whether there are any outliers present, and for comparing distributions. The following example gives an illustration of how to construct a box and whisker plot for 20 repeat measurements of nitrate in drinking water (in mg L^{-1}):

2.3	2.2	2.1	2.4	2.3	2.2	2.9	3.3	3.1	2.3
3.8	2.2	2.4	2.6	2.2	2.1	2.8	2.5	2.1	3.4

- Arrange the results in ascending numerical order and make a tally.

2.1	2.2	2.3	2.4	2.5	2.6	2.7	2.8	2.9	3.0	3.1	3.2	3.3	3.4	3.5	3.6	3.7	3.8
2.1	2.2	2.3	2.4	2.5	2.6		2.8	2.9		3.1		3.3	3.4				3.8
2.1	2.2	2.3	2.4														
2.1	2.2	2.3															
	2.2																

- Calculate the median. This is equal to the value of the central member of the series when there are odd numbers (if there are an even number of values, the median is the mean of the two central values):

Position of median = $(n + 1)/2^{th}$ value = $(20 + 1)/2 = 10.5^{th}$ value = 2.35

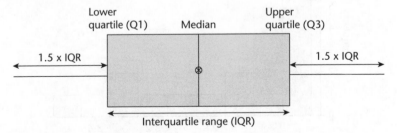

Figure 7.1 Example of a box and whisker plot

When there are an odd number of results, there will be an equal number of data points above and below the median value.

- Calculate the quartiles. There are several ways of doing this, but we shall use the method of calculating the weighted values, which is used in statistical software packages.

Q1 is the value of result number $0.25(n + 1)$, that is, the values of 25% of the results are less than or equal to Q1.

Q3 is the value of result number $0.75(n + 1)$, that is, the values of 75% of the results are more than or equal to Q3.

If the results of $0.25(n + 1)$ or $0.75(n + 1)$ are not integers, the values of Q1 and Q3 must be interpolated.

Q1 = $0.25(20 + 1)$th result = the value of the 5.25th result in the series. This is not an integer so we must interpolate between the 5th and 6th result so:

$$Q1 = 5\text{th result} + 0.25(6\text{th result} - 5\text{th result}) = 2.2 + 0.25(2.2 - 2.2) = 2.2$$

Q3 = $0.75(20 + 1)$th result = the value of the 15.75th result in the series. This is not an integer so we must interpolate between the 15th and 16th result so:

$$Q3 = 15\text{th result} + 0.75(16\text{th result} - 15\text{th result}) = 2.8 + 0.75(2.9 - 2.8) = 2.875$$

- Calculate the IQR.

$$IQR = Q3 - Q1 = 2.875 - 2.2 = 0.675$$

- Draw the whiskers. These are typically set at the lowest or highest datum within 1.5 times the IQR. Data points which are outside the whiskers are usually considered to be outliers. The whiskers are drawn from the edges of the box:

$$Q1 - (1.5 \times IQR) = 2.2 - (1.5 \times 0.675) = 2.2 - 1.0125 = 1.1875$$

$$\text{Lower whisker} = 2.1 \text{ (lowest datum within } 1.5 \times IQR)$$

$$Q3 + (1.5 \times IQR) = 2.875 + (1.5 \times 0.675) = 2.875 + 1.0125 = 3.8875$$

$$\text{Upper whisker} = 3.8 \text{ (highest datum within } 1.5 \times IQR)$$

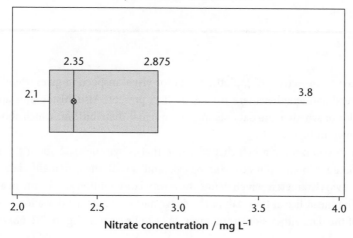

Box plot of nitrate concentration

This example shows that the data is skewed somewhat to higher values because most of the data is higher than the median and the upper whisker extends further than the lower whisker. However, there are no outliers present in this example because all the results fall within the upper and lower whiskers. You should now be able to complete Activity 7.2 using data for nitrate concentration in drinking water, which are close to the regulatory limit.

ACTIVITY 7.2 Visualizing data using box and whisker plots

The results of ten repeat measurements of nitrate in a drinking water, for two different samples, are given in Table 7.1.

Table 7.1 Results of ten replicate determinations of nitrate in drinking water

| Measurement no. | Nitrate concentration/mg L^{-1} | |
	Sample A	Sample B
1	51	52
2	49	51
3	50	49
4	48	48
5	48	49
6	50	50
7	49	51
8	47	52
9	49	51
10	55	50

• Draw a box and whisker plot for each set of data, indicating the median, upper and lower quartiles, the whiskers, and any outliers.

• Comment on the data by comparing the median, spread, and presence of any outliers.

• When you have completed this activity, study the feedback given at the end of the chapter.

As can be seen from Activity 7.2, plotting data for visual inspection gives useful information about its spread and whether there are any outliers present. Also, histograms and blob plots give some idea of whether the data belong to a **normal distribution**, which should look like the curve shown in Figure 7.2.

A normal distribution is a bell-shaped curve that is symmetrical about the mean and is important because many physical, biological, and social measurement data exhibit this distribution. Two parameters which define the normal distribution are the **mean** and **standard deviation**. The mean lies at the peak of the curve and the standard deviation is a measure of the spread of the data either side of the mean. As can be seen in Figure 7.2, 68.3% of the data

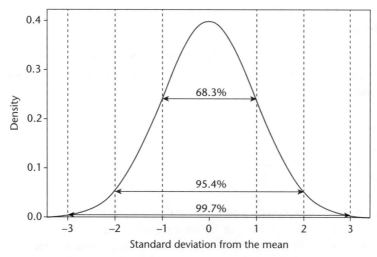

Figure 7.2 A normal distribution showing the approximate proportion of the dataset accounted for by $\pm 1s$ (68.3%), $\pm 2s$ (95.4%), and $\pm 3s$ (99.7%) either side of the mean

lie within one standard deviation either side of the mean ($\pm 1s$), 95.4% within $\pm 2s$, and 99.7% within $\pm 3s$. This is an important observation which will be discussed in more detail in the following section. Most statistical tests assume a normal distribution, an assumption which is made in this book. However, it is worth noting that sometimes data will follow a non-normal distribution where the curve shown in Figure 7.2 will look skewed to one side and what may look like an outlier could well be a valid piece of information.

Visual inspection is a useful way of estimating the spread in data and identifying differences between different sets of data. However, it is a rather subjective way of comparing measurements, so in order to be more scientifically rigorous we need to use some basic statistical tests.

. . . .

7.3 **Statistical calculations**

Most statistics assume a normal distribution so, if we are confident that the data is normally distributed, we can use a range of **statistical tests** to compare measurements objectively. There is often much confusion surrounding the use of such tests. Commonly-asked questions include: when and how they should be applied; which one should be used; and what do they actually show? Hence it is first necessary to understand some basic principles before proceeding.

7.3.1 **Population and sample**

In statistics the **population** includes all possible measurements of a particular parameter. For example, if you wanted to calculate the average height of males in the UK it would be necessary to measure the height of every male individual before you could say that you had sampled the population. In practice, it is more practical to measure a **sample** of the population, that is, a limited number of measurements which we hope will represent the population. Likewise in analytical science, it is most often impractical or impossible to make a large enough number of analytical measurements to approximate the *whole* population so we normally have to make do with a sample. The sample size will vary depending on the practicalities of the measurement being made.

7.3.2 Mean

You can easily calculate the arithmetic mean (or average) using a calculator or spreadsheet. Add the individual results together, and divide the total by the number of results. Mathematically this is expressed as:

Population mean $\qquad \mu = \dfrac{\sum_{i=1}^{N} x_i}{N}$ $\qquad\qquad\qquad\qquad$ (7.1)

Sample mean $\qquad\quad \bar{x} = \dfrac{\sum_{i=1}^{n} x_i}{n}$ $\qquad\qquad\qquad\qquad$ (7.2)

In this case any result is represented as x_i, where i can be any integer value from 1 to n, and n is the total number of results. Most of the time we only make a relatively small number of measurements (n is small) so $\bar{x}$ is the **sample mean**. If we are able to make the maximum number of measurements possible, N, then the value calculated in this way is called the **population mean**, μ, often referred to as the **true value**. As the number of results n increases, then $\bar{x}$ gets closer to μ.

The main point to understand from this is that we can normally only make a limited number of measurements, so the sample mean will only be an estimate of the population mean or true value.

7.3.3 Variance

The variance is a measure of the spread of data about the mean and is the square of the standard deviation (see Section 7.3.4). More correctly, the standard deviation is the square root of the variance because the variance must be calculated first; however, because the standard deviation is used most frequently for describing data and electronic calculators allow you to calculate it directly, this important point is often overlooked.

Population variance $\qquad \sigma^2 = \dfrac{\sum_{i=1}^{N} (x_i - \mu)^2}{N}$ $\qquad\qquad\qquad$ (7.3)

Sample variance $\qquad\quad s^2 = \dfrac{\sum_{i=1}^{n} (x_i - \bar{x})^2}{n-1}$ $\qquad\qquad\qquad$ (7.4)

Note the important difference between the population and sample variance; the divisor is N in the former and $n-1$ in the latter. There is a reason for this, as explained in the next section.

7.3.4 Standard deviation

The standard deviation is also a measure of the spread of data about the mean, and is the most frequently used statistical measure. You can use a calculator or spreadsheet to calculate this, but you must be sure that you are calculating the correct standard deviation. On a scientific calculator there are two versions of standard deviation, with the symbols σ_n and σ_{n-1}. These are the population and sample standard deviation, respectively.

Population standard deviation $\qquad \sigma_n = \sigma = \sqrt{\dfrac{\sum_{i=1}^{N} (x_i - \mu)^2}{N}}$ $\qquad\qquad$ (7.5)

Sample standard deviation $\qquad \sigma_{n-1} = s = \sqrt{\dfrac{\sum_{i=1}^{n}\left(x_i - \overline{x}\right)^2}{n-1}}$ $\qquad$ (7.6)

Under most circumstances you will need to use the sample standard deviation, s. For example, if you are asked to determine the concentration of mercury in Atlantic tuna fish on a particular day then you would most likely take a sample of tuna fish, determine the concentration of mercury in each and calculate the sample mean and standard deviation. You would clearly be unable to sample the whole population of tuna from the Atlantic Ocean so your sample mean and standard deviation will be an estimate of the population mean and standard deviation.

When calculating the standard deviation for a small sample size it is quite easy to underestimate its magnitude compared to the population, so the divisor is set as $n-1$ in order to give a slight overestimate to account for this. Under almost all circumstances you should use the sample standard deviation, s.

Distributions of two sets of data are shown visually in Figure 7.3, where two different datasets have been plotted. These represent the concentration of NO_3^- in water at two different drinking water treatment plants, and shows that the mean and spread of measurements for Plant 1 ($\overline{x} = 4.355$, $s = 3.344$) is greater than for Plant 2 ($\overline{x} = 3.135$, $s = 2.118$).

7.3.5 Relative standard deviation

The relative standard deviation (RSD) is the ratio of the standard deviation to the mean:

$$s_r = \frac{s}{\overline{x}}$$ $\qquad$ (7.7)

usually expressed as a percentage:

$$s_r = \frac{s}{\overline{x}} \times 100\%$$ $\qquad$ (7.8)

This is useful when you wish to compare the spread of data for measurements with different means, particularly when they are of quite different magnitude.

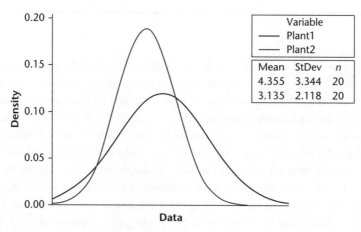

Figure 7.3 Distributions of two separate datasets, showing the spread of data

ACTIVITY 7.3 Mean and standard deviation

Calculate the mean, sample standard deviation, and RSD for nitrate concentration in drinking water for samples A and B given in Table 7.1 in Activity 7.2.

Sample standard deviation $\sigma_{n-1} = s = \sqrt{\dfrac{\sum_{i=1}^{n}\left(x_i - \bar{x}\right)^2}{n-1}}$

Data table for standard deviation:

x_i	$x_i - \bar{x}$	$\left(x_i - \bar{x}\right)^2$
Sums:		

When you have completed this activity, study the feedback given at the end of this chapter.

Take another look at the box and whisker plots from Activity 7.2 and the means you have calculated in Activity 7.3. Would you say that the mean nitrate concentration in sample A was different from the mean nitrate concentration in sample B? If, in answer to this question you decided that the mean nitrate concentration in sample B is higher than the mean nitrate concentration in sample A then, on the face of it, you would be correct. However, look at the spread of the data shown in the box and whisker plots; although the means are different there is considerable **overlap** between the data, so the decision is not quite as clear-cut as it first seems. In order to make an objective decision as to whether there is any **significant** difference between the means we need to apply a **statistical test**. In statistics a significant difference is one which can be stated with a certain degree of **confidence** that it is not due to **random error**.

7.3.6 Standard error of the mean

One more definition is required before we can discuss the concept of a statistical test. The standard error of the mean (s.e.m.) is the standard deviation of the sampling distribution of the sample means. If you took repeat samples from a population, calculated their means and plotted the values, the distribution of the means would look more normal than the population distribution. The standard error of the mean, which is the standard deviation of the sample means, can be estimated from the standard deviation of the original as follows:

$$\text{Standard error of the mean} \qquad \text{s.e.m.} = \frac{s}{\sqrt{n}} \qquad\qquad (7.9)$$

where s is the sample standard deviation of the individual means and n is the sample size. This equation is an estimate of the standard deviation of the means ($s_{\bar{x}}$) which is approximately correct as long as the sample size is less than 1/20th of the population size.

7.3.7 Degrees of freedom

The number of **degrees of freedom** (d.f.) that you should use when estimating a statistical parameter is equal to the number of independent measurements that go into the estimate. For example, in the calculation of the mean all the measurements are independent so the d.f. $= n$, however, when calculating the sample standard deviation the mean must first be calculated which reduces the number of degrees of freedom by one, so d.f. $= n - 1$. Don't worry about this because there are only a limited number of instances in which you will have to remember how many degrees of freedom should be used for a particular statistical test.

7.4 Statistical tests

Up to now we have represented data in different ways and performed some basic calculations to see if we can spot differences between different sets of data. However, while the brain is very good at recognizing differences between patterns there are some instances when we need a more objective approach on which we can base our decisions. This is where statistical tests come in, because they provide an objective way of comparing sets of data to evaluate **statistically significant** differences which are not simply down to chance. Nearly all statistical tests are performed on **samples** of the population, so we need to use parameters for samples rather than the population. This is an important point to which we will come back.

Statistical tests are not absolute, but are based on **probability**: they test the truth of a hypothesis called the **null hypothesis (H$_0$)** which states that there is no difference between population parameters. So, we start off by assuming that there is no difference and then use a test to calculate the **probability** (p) of rejecting the null hypothesis when it is true. For example, if $p = 0.05$ (or 5%) there is a 5% chance of rejecting the null hypothesis when it is in fact true, and if $p = 0.01$ then there is only a 1% chance. So the smaller the value of p the more likely it is that that there is a statistically significant difference between two values which is not caused by chance.

In practice, we set a value of p at which we choose to reject the null hypothesis with a given degree of **confidence** and, in effect, decide that there is a significant difference between the observed values. It is usual to say that differences in data with values of $p \leq 0.05$ (a confidence level of 95%) are statistically significant, that is, that the observed difference is unlikely to happen by chance. If we set the confidence level at 95% ($p \leq 0.05$) then we come to the wrong decision 5% of the time. This is a so-called **Type 1** error (i.e. rejecting the null hypothesis when it is in fact true). In some situations this may not be good enough and we may wish to use a higher level of confidence.

$p \leq 0.05$ 95% confidence level; wrong decision 5% of the time

$p \leq 0.01$ 99% confidence level; wrong decision 1% of the time

$p \leq 0.001$ 99.9% confidence level; wrong decision 0.1% of the time

Statistical tests are useful because they allow you to compare sets of data in an objective way. For example, you may want to compare two methods for the determination of lead in drinking water, or to compare a new method with an old method. There are two types of statistical tests which you will find useful in analytical science, the **t-test** and **F-test.**

7.4.1 The *t*-test

A *t*-test is a statistical test used to test whether a sample mean equals a hypothesized value (e.g. to compare a measured value with a known standard), or to compare two sampling means (e.g. to compare two analytical methods). There are two versions:

One sided: used to determine if a measured result is different from a known value, but only in one direction. For this test it must be decided beforehand in which direction, either greater than or less than, for example, to decide whether a measured value for nitrate in drinking water is above the legislative limit.

Two sided: used to determine if a measured result is different in any direction. For this test no prior decision must be made about the direction, for example, to decide whether method A gives a different result to method B, either greater than or less than.

The difference between a one-sided and two-sided test will become more apparent when you work through the activities.

The calculation for a *t*-test is straightforward, but you will need to look up the critical value of *t* in a set of statistical tables (see Appendix A, Table A1). Alternatively, you can use a calculator or the statistical functions in a spreadsheet. Regardless of this, you must use the correct formula for the test statistic as follows:

To test whether the population mean equals a hypothesized value, for example, to compare the measured and certified values of a reference material:

$$t = \frac{\bar{x} - \mu}{s/\sqrt{n}} \tag{7.10}$$

where $\bar{x}$ = sample mean, μ = true (or hypothesized) value, s = sample standard deviation, and n = sample size.

For comparison of two sample means, for example, to compare the results of two analytical methods:

$$t = \frac{\bar{x}_1 - \bar{x}_2}{s_c \sqrt{\dfrac{1}{n_1} + \dfrac{1}{n_2}}} \tag{7.11}$$

where $\bar{x}_1$ = mean of sample 1, $\bar{x}_2$ = mean of sample 2, n_1 = size of sample 1, n_2 = size of sample 2, and s_c = pooled standard deviation (defined on page 173).

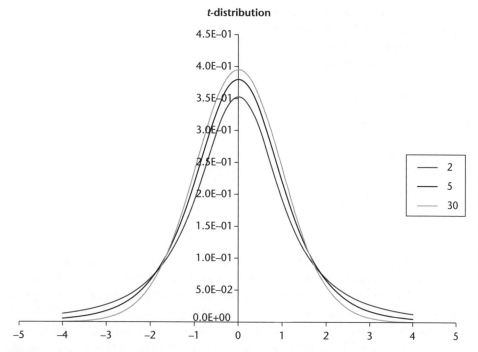

Figure 7.4 The *t*-distribution showing how the wings narrow as the number of degrees of freedom increases from 2 to 30

In the second case, because both sets of measurements have a mean and standard deviation associated with them, and providing that there is no significant difference between their standard deviations (see *F*-test later), we need to calculate the pooled standard deviation s_c:

$$s_c = \sqrt{\frac{s_1^2(n_1 - 1) + s_2^2(n_2 - 1)}{(n_1 + n_2 - 2)}} \qquad (7.12)$$

where s_1 = standard deviation of sample 1 and s_2 = standard deviation of sample 2.

It is important to realize that the *t*-test is designed to be used on samples rather than the whole population. The reason for this is because it is based on a probability distribution called the **t-distribution** which is similar to the standard normal distribution but takes into account the effect of small sample sizes for $n < 30$. This spreads the distribution out so that the *t*-distribution is fatter in the tails compared to the normal distribution, but it tends to the normal distribution as *n* increases (Figure 7.4).

E XAMPLE 7.1

Performing a *t*-test to compare a mean with a true value

You have determined the concentration of nitrate in the certified reference material (CRM) RTC-QCI-028K (certified value 12.4 mg L^{-1} undiluted water) and obtained a mean result for the measured value $\bar{x}$ = 11.6 mg L^{-1} and a standard deviation s = 0.96 mg L^{-1} for five replicate determinations. Use a two-sided test to compare your measured value ($\bar{x}$) for nitrate in the CRM with the certified value (μ).

Null hypothesis: there is no difference between the measured mean and certified values other than that caused by random error.

Measured $\bar{x} = 11.6$ mg L^{-1}

$s = 0.96$ mg L^{-1}

$n = 5$

Certified $\mu = 12.4$ mg L^{-1}

Calculate the value of t:

$$t = \frac{\bar{x} - \mu}{s/\sqrt{n}}$$

$$t = \frac{11.6 - 12.4}{0.96/\sqrt{5}} = -1.86$$

$$t_{calc} = -1.86$$

Before we go further it is worth considering what you have calculated. The t_{calc} value is the difference between the sample mean ($\bar{x}$) and the population mean (μ) divided by $s/\sqrt{n}$, which is the same as the estimated standard error of the mean (s.e.m.). So t_{calc} is a measure of how far away the sample mean is from the population mean **in multiples of the s.e.m.** We can compare this value with a table of critical values of t (t_{crit}) arranged at different levels of confidence and degrees of freedom. Before doing this comparison, let us look at part of Table A1 given in Appendix A and consider its structure:

d.f.						
One sided	0.90	0.95	0.975	0.995	0.9975	0.9995
Two sided	0.80	0.90	0.95	0.99	0.995	0.999
1	3.078	6.314	12.706	63.657	127.321	639.619
2	1.886	2.920	4.303	9.925	14.089	31.598
3	1.638	2.353	3.182	5.841	7.453	12.941
4	1.533	2.132	2.776	4.604	5.598	8.610
5	1.476	2.015	2.571	4.032	4.773	6.859

The main body of the table contains the critical value of t for various levels of confidence (columns) and degrees of freedom (rows).

The first thing to note is that all the values of t_{crit} are positive. This is because the table only shows values for the positive half of the t-distribution (the right hand side in Figure 7.4). Our value of t_{calc} was negative because of the way we arranged the calculation (i.e. the difference between the sample and population means in the negative direction). Because this is a two-sided test we are considering the difference in either direction, both positive and negative, so we can ignore the sign and take the absolute value:

Hence: $\left| t_{calc} \right| = 1.86$

We need to compare our value of t_{calc} with the appropriate value of t_{crit} for a specific number of degrees of freedom and at our chosen level of confidence. This provides a threshold value at which we can make a decision about the null hypothesis:

Level of confidence: $p = 0.05$ (95%)

Degrees of freedom: d.f. $= n - 1 = 4$

In this case the number of degrees of freedom (d.f.) $= n - 1$. We have reduced the number of independent measurements that go into the estimate because it is necessary to use the standard deviation, for which the mean must be first calculated.

$$\left|t_{calc}\right| = 1.86$$
$$t_{crit} = 2.776$$

Hence, we can reject H_0 if $\left|t_{calc}\right| \geq t_{crit}$ with only a 5% chance of being wrong.

In this case $\left|t_{calc}\right| < t_{crit}$ so we can retain the null hypothesis and say that there is no statistically significant difference between the measured mean and the certified value at the $p = 0.05$ (or 95% confidence) level.

An easier way of visualizing this is by using the t-distribution itself, as shown in Figure 7.5. The arrows indicate the critical value of $t = 2.776$ for 4 d.f. and 95% confidence which we have read from Table A1 in Appendix A. Between these values, the curve encapsulates 95% of the distribution, while 2.5% lies above $t = 2.776$ and 2.5% lies below $t = -2.776$. It is apparent then that 5% of the total distribution lies above or below the critical value of t, so if our calculated value lies between $t = 2.776$ and $t = -2.776$ then we can retain the null hypothesis; but if it lies outside in the tails then we will reject it erroneously only 5% of the time. In practice we ignore the sign and just check to see if $t_{calc} > 2.776$ and this takes care of both positive and negative directions.

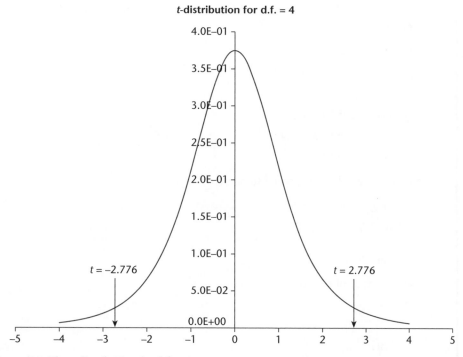

Figure 7.5 The t-distribution for d.f. $= 4$

EXAMPLE 7.2

Performing a *t*-test to compare two means

You have also measured the nitrate concentration in two samples of drinking water and obtained the data given in Table 7.2. You wish to compare the samples with each other to see if they are different and also to compare each sample to the regulatory limit of 50 mg L^{-1}.

Table 7.2 Results of ten replicate determinations of nitrate in drinking water

| Measurement no. | Nitrate concentration/mg L^{-1} | |
	Sample A	Sample B
1	51	52
2	49	51
3	50	49
4	48	48
5	48	49
6	50	50
7	49	51
8	47	52
9	49	51
10	55	50

- Two-sided test—compare sample A with sample B to see if there is a significant difference.

First, you need to calculate the mean ($\bar{x}$) and sample standard deviation (*s*) for each of the samples:

Sample A: $\bar{x}_1 = 49.6$ mg L^{-1}

$s_1 = 2.2211$ mg L^{-1}

$n_1 = 10$

Sample B: $\bar{x}_2 = 50.3$ mg L^{-1}

$s_2 = 1.3375$ mg L^{-1}

$n_2 = 10$

Next you must calculate the pooled standard deviation, s_c:

$$s_c = \sqrt{\frac{s_1^2(n_1-1)+s_2^2(n_2-1)}{(n_1+n_2-2)}}$$

$s_c = 1.8336$ mg L^{-1}

Calculate the value of *t*:

$$t = \frac{\bar{x}_1 - \bar{x}_2}{s_c\sqrt{\frac{1}{n_1}+\frac{1}{n_2}}}$$

$t_{calc} = -0.854$

Because this is a two-sided test we want to know whether the two methods differ in any direction so we can ignore the sign of t and take the absolute value:

$$\left| t_{calc} \right| = 0.854$$

The null hypothesis is that there is no difference between the measured mean concentrations in sample A and B.

H_0: $\bar{x}_1 = \bar{x}_2$, which we can reject if $\left| t_{calc} \right| \geq t_{crit}$.

The critical value of t, t_{crit}, must be found from statistical tables and will depend on the level of confidence and the number of degrees of freedom. In this case d.f. $= n_1 + n_2 - 2$ because two parameters (the means, $\bar{x}_1$ and $\bar{x}_2$, used to calculate s_1 and s_2) have been used in intermediate steps.

Level of confidence, $p = 0.05$
Degrees of freedom: d.f. $= n_1 + n_2 - 2 = 18$
Hence: $t_{crit} = 2.1$

In this case $\left| t_{calc} \right| < t_{crit}$ so we can retain the null hypothesis and say that there is no significant difference between the measured means of samples A and B at the 95% confidence level.

- One-sided test—compare the concentration of nitrate in each of the samples to the regulatory limit of 50 mg L^{-1}.

In a one-sided test there is an added complication of 'sign' or 'direction' of comparison which we must be aware of when using the t-value. The null hypothesis is that there is no difference between the sample mean and the regulatory limit, μ, *other than due to random error*.

Case 1: we want to know whether the means of sample A and B **exceed** the regulatory limit, an example of a **one-sided** test where we really want to know if $\bar{x} > \mu$.
Calculate the value of t for each of the samples:

$$t = \frac{\bar{x} - \mu}{s / \sqrt{n}}$$

Sample A: $\bar{x} = 49.6$ mg L^{-1}
$s = 2.2211$ mg L^{-1}
$n = 10$
$t_{calc} = -0.569$

Sample B: $\bar{x} = 50.3$ mg L^{-1}
$s = 1.3375$ mg L^{-1}
$n = 10$
$t_{calc} = 0.709$

In this case we need the one-sided value of t_{crit} for a 95% level of confidence ($p = 0.05$) and $(n - 1)$ degrees of freedom, obtained from the t-distribution in Appendix A (note that you must use a different column to the two-sided test).
For samples A and B:

d.f. $= 9$
$t_{crit} = 1.833$

In this case $t_{calc} < t_{crit}$ so for both sample A and B neither sample exceeds the regulatory limit.

Note: This shows an important result: even though the mean for sample B is greater than the regulatory limit, there is no statistically significant difference because there is a spread of results either side of the regulatory limit.

Case 2: we want to know whether the means of sample A and B are **less than** the regulatory limit, an example of a **one-sided** test where we really want to know if $\bar{x} < \mu$.

This is where a complication arises because for an 'is it less than' comparison we need to use the negative value of t_{crit} which is -1.833, so from the table of critical values of t in Appendix A, $-t_{crit} = -1.833$.

This is most easily explained using the t-distribution shown in Figure 7.6. If we want to know whether the measured value is more than the regulatory limit we use the right hand side of the t-distribution, where values of t are positive. However, if we want to know if the measured value is less than the regulatory limit we must use the left hand side of the t-distribution where values of t are negative.

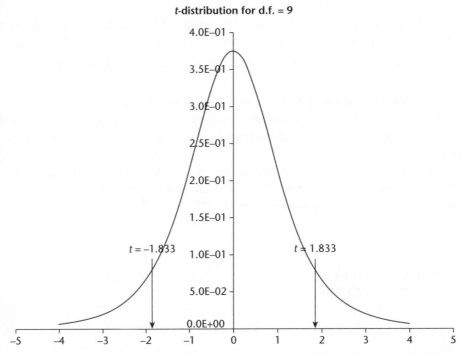

Figure 7.6 The t-distribution for d.f. = 9

Sample A: $t_{calc} = -0.569$ and $-t_{crit} = -1.833$

In this case $t_{calc} > -t_{crit}$ so the mean for sample A is not less than the regulatory limit and the null hypothesis is retained.

Sample B: $t_{calc} = 0.709$ and $-t_{crit} = -1.833$

In this case $t_{calc} > -t_{crit}$ so the mean for sample B is not less than the regulatory limit and the null hypothesis is retained.

Note: This does not mean that samples A and B are greater than the regulatory limit, just that the means are not significantly different from it.

In summary, we can say that samples A and B are neither greater than nor less than the regulatory limit and that there is no significant statistical difference between them, or between their respective means and the regulatory limit.

When comparing the means of samples the two cases just discussed illustrate the difference between a one-sided and two-sided test. Using these examples you can see that the same critical value of *t* is used for both a one- and two-sided test but with a **different level of confidence**:

For 9 d.f. and $t_{crit} = 1.833$
One-sided test   confidence = 95%
Two-sided test   confidence = 90%

This is because, for a two-sided test we are checking to see whether t_{calc} is outside the range 1.833 to −1.833, which contains 90% of the distribution, with 10% outside in the tails. For a one-sided test we are only using one side of the distribution so must take account of the sign depending on the comparison we wish to make.

7.4.2 The *F*-test

In the preceding examples, when we have compared means of two different samples using the *t*-test, we have assumed that there is no significant difference between their variances, that is, that there is similar spread in the data. If there is a significant difference in the spread of data it is not possible to combine the variances and the *t*-test becomes invalid, so it is necessary to check this beforehand. We can do this using the *F*-test, which is simply a ratio of the variances of the two sets of data being compared.

$$F = \frac{s_1^2}{s_2^2} \text{ where } s_1 > s_2 \tag{7.13}$$

Clearly, if the variances are the same then *F* = 1. For reasons of simplicity the formula is arranged so that the numerator is greater than the denominator, so that the variances are non-equal when *F* > 1. We need to test the significance of this so, much like a *t*-test, a table of values for the distribution are used to look up the critical value of *F* for a given level of confidence and number of degrees of freedom for the numerator and denominator using Table A2 in Appendix A. If $F_{calc} > F_{crit}$ then there is a significant difference.

EXAMPLE 7.3

Performing an *F*-test to combine standard deviations
Using the data in Table 7.2:

Sample A: $\bar{x}_1 = 49.6$ mg L^{-1}
            $s_1 = 2.2211$ mg L^{-1}
            $s_1^2 = 4.93$
            $n_1 = 10$
            d.f. = 9

Sample B: $\bar{x}_2 = 50.3$ mg L^{-1}
            $s_2 = 1.3325$ mg L^{-1}
            $s_2^2 = 1.80$

$$n_2 = 10$$
$$\text{d.f.} = 9$$
$$F_{calc} = \frac{s_1^2}{s_2^2} = \frac{4.93}{1.80} = 2.76$$

We now need to look up the critical value of F in tables for a given level of confidence and taking into account the degrees of freedom for both samples. In this case we are performing a two-sided test so you should make sure that you are using the correct set of tables; Table A3 in Appendix A is for two-sided critical values of F.

For a 95% confidence level and 9, 9 d.f.

$$F_{crit} = 4.03$$

In this case $F_{calc} < F_{crit}$ so there is no significant difference between the variances, so we can combine them for the t-test.

PROBLEM 7.1

Determination of nitrate in drinking water

You have just started a new job as quality control manager of a water treatment works. As one of your first jobs you have been asked to perform the following tasks:

- to determine whether the new method gives results which are significantly different from the old method;
- to check the method for systematic error;
- to compare the mean concentration of nitrate in drinking water from two different treatment plants measured over a 20 day period;
- to check whether the mean concentration of nitrate in drinking water, measured over the 20 day period from Plant 1, exceeds a proposed new regulatory limit of 10 mg L⁻¹.

The results that you need to analyse are given in Tables 7.3–7.5.

Table 7.3 Comparison of two methods for nitrate in drinking water

	Concentration of nitrate/mg L⁻¹					Mean	s
Old method	4.20	3.50	4.80	5.20	4.35	4.41	0.643
New method	3.55	3.10	2.95	3.50	3.65	3.35	0.306

Table 7.4 Replicate analysis of a quality control sample over 1 month using the new method

Certified nitrate concentration/mg L⁻¹	Mean measured nitrate concentration/mg L⁻¹	s	n
3.54	3.64	±0.27	30

Table 7.5 Data for the concentration of nitrate in drinking water determined over a period of 20 days at two different water treatment plants

| Day | Nitrate concentration/mg L^{-1} | |
	Plant 1	Plant 2
1	3.1	6.0
2	2.5	5.6
3	1.3	2.5
4	4.1	1.2
5	3.4	7.1
6	2.3	2.7
7	9.2	2.2
8	10.9	1.1
9	10.1	2.1
10	3.1	1.2
11	1.8	1.1
12	2.1	3.1
13	4.2	1.1
14	6.4	3.7
15	2.6	2.3
16	1.2	3.2
17	8.1	2.5
18	5.8	1.5
19	1.5	4.1
20	1.4	8.4

Feedback on activities and problems

 FEEDBACK ON ACTIVITY 7.1

Finding regulatory information on drinking water

The Drinking Water Inspectorate website (http://www.dwi.gov.uk/) gives advice to consumers on water quality. You should be able to download a consumer advice leaflet at:

http://www.dwi.gov.uk/consumers/advice-leaflets/index.htm.

At the time of publication, the MAC for nitrate in drinking water in the UK is 50 mg L^{-1}.

 FEEDBACK ON ACTIVITY 7.2

Visualizing data using box and whisker plots

- *Draw a box and whisker plot for each set of data, indicating the median, upper and lower quartiles, the whiskers, and any outliers.*

Sample A

Arrange the data in order on a horizontal scale:

47	48	49	50	51	52	53	54	55
47	48	49	50	51				55
	48	49	50					
		49						

Position of median = $(n + 1)/2$th value = $(10 + 1)/2 = 5.5$th value = 49

Lower quartile, Q1 = $0.25(10 + 1)$th result = value of the 2.75th result = 48

Upper quartile, Q3 = $0.75(10 + 1)$th result = value of the 8.25th result = 50.25

IQR = Q3 − Q1 = 50.25 − 48 = 2.25

Q1 − $(1.5 \times IQR)$ = 48 − (1.5×2.25) = 48 − 3.375 = 44.625

Lower whisker = 47

Q3 + $(1.5 \times IQR)$ = 50.25 + (1.5×2.25) = 50.25 + 3.375 = 53.625

Upper whisker = 51

Sample B

Arrange the data in order on a horizontal scale:

47	48	49	50	51	52	53	54	55
	48	49	50	51	52			
		49	50	51	52			
				51				

Position of median = $(n + 1)/2$th value = $(10 + 1)/2 = 5.5$th value = 50.5

Lower quartile, Q1 = $0.25(10 + 1)$th result = value of the 2.75th result = 49

Upper quartile, Q3 = $0.75(10 + 1)$th result = value of the 8.25th result = 51.25

IQR = Q3 − Q1 = 51.25 − 49 = 2.25

Q1 − $(1.5 \times IQR)$ = 49 − (1.5×2.25) = 49 − 3.375 = 45.625

Lower whisker = 48

Q3 + $(1.5 \times IQR)$ = 51.25 + (1.5×2.25) = 51.25 + 3.375 = 54.625

Upper whisker = 52

50.5

- *Comment on the data by comparing the median, spread, and presence of any outliers.*

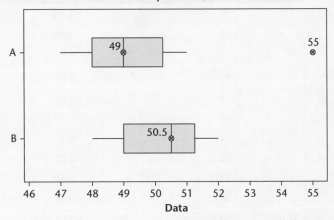

Box plots of A, B

Comparison of the box and whisker plots above reveals that the median for sample A is lower than for sample B; however, there is considerable overlap of their IQRs. Sample A has an outlier value at 55. Sample B exhibits some skew in the distribution of data to the higher end.

 FEEDBACK ON ACTIVITY 7.3

Mean and standard deviation

Calculate the mean, sample standard deviation, and RSD for nitrate concentration in drinking water for samples A and B given in Table 7.1 in Activity 7.2.

Data tables for standard deviation:

	Sample A		
	x_i	$x_i - \bar{x}$	$\left(x_i - \bar{x}\right)^2$
	51	1.4	1.96
	49	−0.6	0.36
	50	0.4	0.16
	48	−1.6	2.56
	48	−1.6	2.56
	50	0.4	0.16
	49	−0.6	0.36
	47	−2.6	6.76
	49	−0.6	0.36
	55	5.4	29.16
Sums:	496	0	44.4

	Sample B		
	x_i	$x_i - \bar{x}$	$(x_i - \bar{x})^2$
	52	1.7	2.89
	51	0.7	0.49
	49	−1.3	1.69
	48	−2.3	5.29
	49	−1.3	1.69
	50	−0.3	0.09
	51	0.7	0.49
	52	1.7	2.89
	51	0.7	0.49
	50	−0.3	0.09
Sums:	503	0	16.1

Sample A

Mean:

$$\bar{x} = \frac{496}{10} = 49.6$$

Sample standard deviation: $s = \sqrt{\dfrac{\sum_{i=1}^{n}(x_i - \bar{x})^2}{n-1}} = \sqrt{\dfrac{44.4}{(10-1)}} = 2.2211$

Relative standard deviation: $RSD = \dfrac{s}{\bar{x}} \times 100 = \dfrac{2.22}{49.6} \times 100 = 4.5\%$

Sample B

Mean:

$$\bar{x} = \frac{503}{10} = 50.3$$

Sample standard deviation: $s = \sqrt{\dfrac{\sum_{i=1}^{n}(x_i - \bar{x})^2}{n-1}} = \sqrt{\dfrac{16.1}{(10-1)}} = 1.3375$

Relative standard deviation: $RSD = \dfrac{s}{\bar{x}} \times 100 = \dfrac{1.34}{50.3} \times 100 = 2.7\%$

? FEEDBACK ON PROBLEM 7.1

Determination of nitrate in drinking water

You have just started a new job as quality control manager of a water treatment works. As one of your first jobs you have been asked to perform the following tasks:

- *to determine whether the new method gives results which are significantly different from the old method;*

You can do this in two ways, using a *t*-test and also with box and whisker plots.

	Concentration of nitrate/mg L^{-1}					Mean	s	n
Old method	4.20	3.50	4.80	5.20	4.35	4.41	0.643	5
New method	3.55	3.10	2.95	3.50	3.65	3.35	0.306	5

You can compare the mean concentration of nitrate determined using the two methods using a t-test, but first you must check whether it is possible to combine the variances to calculate the pooled standard deviation.

F-test: $F_{calc} = \dfrac{s_1^2}{s_2^2} = \dfrac{(0.643)^2}{(0.306)^2} = 4.42$

For a 95% confidence level and 4, 4 d.f., and using Table A3 in Appendix A for a two-sided test,

$F_{crit} = 9.60$

$F_{calc} < F_{crit}$ so there is no significant difference between the variances, so we can combine them for the t-test.

Pooled standard deviation: $s_c = \sqrt{\dfrac{s_1^2(n_1-1)+s_2^2(n_2-1)}{(n_1+n_2-2)}} = \sqrt{\dfrac{1.6538+0.3745}{8}}$

$s_c = 0.5035$

t-test: $t_{calc} = \dfrac{\bar{x}_1 - \bar{x}_2}{s_c\sqrt{\dfrac{1}{n_1}+\dfrac{1}{n_2}}} = \dfrac{1.06}{0.3184} = 3.33$

Level of confidence: 95% ($p = 0.05$)

Degrees of freedom: d.f. $= n_1 + n_2 - 2 = 8$

Two-sided test: $t_{crit} = 2.306$

$t_{calc} > t_{crit}$ so the mean result for the new method differs significantly from the old method.

You can also use box and whisker plots to compare the median and spread of data, as shown below. This reveals that the median value for the new method is much lower than the old method, almost falling below the end of the lower whisker. The IQRs do not overlap at all.

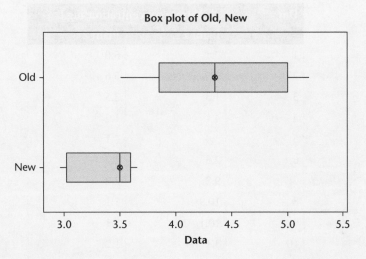

- *to check the method for systematic error;*
You can do this using a t-test.

Certified nitrate concentration /mg L⁻¹	Mean measured nitrate concentration /mg L⁻¹	s	n
3.54	3.64	±0.27	30

Null hypothesis: there is no difference between the measured mean and certified values other than due to random error:

Measured $\bar{x}$ = 3.64 mg L⁻¹
s = 0.27 mg L⁻¹
n = 30
Certified μ = 3.54 mg L⁻¹

Calculate the value of t:

$$t = \frac{\bar{x} - \mu}{s/\sqrt{n}}$$

$$t_{calc} = \frac{3.64 - 3.54}{0.27/\sqrt{30}} = 2.03$$

$$t_{calc} = 2.03$$

Level of confidence: 95% (p = 0.05)
Degrees of freedom: d.f. = $n - 1$ = 29
Two-sided test: t_{crit} = 2.045

$t_{calc} < t_{crit}$ so the measured value does not differ significantly from the certified value.

- *to compare the mean concentration of nitrate in drinking water from two different treatment plants measured over a 20 day period;*

You can do this in two ways, using a t-test and also with box and whisker plots.

Day	Nitrate concentration/mg L⁻¹	
	Plant 1	Plant 2
1	3.1	6.0
2	2.5	5.6
3	1.3	2.5
4	4.1	1.2
5	3.4	7.1
6	2.3	2.7
7	9.2	2.2
8	10.9	1.1
9	10.1	2.1
10	3.1	1.2
11	1.8	1.1
12	2.1	3.1

Day	Nitrate concentration/mg L^{-1}	
	Plant 1	Plant 2
13	4.2	1.1
14	6.4	3.7
15	2.6	2.3
16	1.2	3.2
17	8.1	2.5
18	5.8	1.5
19	1.5	4.1
20	1.4	8.4

You can compare the mean concentration of nitrate determined using the two methods using a t-test, but first you must check whether it is possible to combine the variances to calculate the pooled standard deviation.

Plant 1: $\bar{x}_1 = 4.255$ mg L^{-1}
$s_1 = 3.096$ mg L^{-1}
$n_1 = 20$

Plant 2: $\bar{x}_2 = 3.135$ mg L^{-1}
$s_2 = 2.118$ mg L^{-1}
$n_2 = 20$

F-test: $F_{calc} = \dfrac{s_1^2}{s_2^2} = \dfrac{(3.096)^2}{(2.118)^2} = 2.14$

For a 95% confidence level and 19, 19 d.f., and using Table A3 in Appendix A for a two-sided test,

$$F_{crit} = 2.53$$

$F_{calc} < F_{crit}$ so there is no significant difference between the variances, so we can combine them for the t-test.

Pooled standard deviation: $s_c = \sqrt{\dfrac{s_1^2(n_1-1)+s_2^2(n_2-1)}{(n_1+n_2-2)}} = \sqrt{\dfrac{182.1+85.23}{38}}$

$s_c = 2.6525$

t-test: $t_{calc} = \dfrac{\bar{x}_1 - \bar{x}_2}{s_c\sqrt{\dfrac{1}{n_1}+\dfrac{1}{n_2}}} = \dfrac{1.12}{0.839} = 1.33$

Level of confidence: 95% ($p = 0.05$)
Degrees of freedom: d.f. $= n_1 + n_2 - 2 = 38$
Two-sided test: $t_{crit} = 2.024$

$t_{calc} < t_{crit}$, so there is no significant difference between the mean nitrate concentrations in the two plants.

You can also use box and whisker plots to compare the median and spread of data.

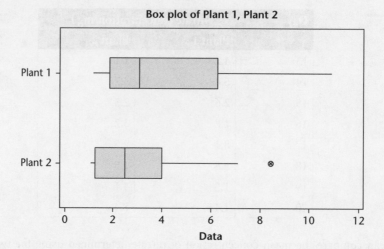

This is useful because it shows that there is an outlier associated with the Plant 2 data and high concentrations of nitrate on days 7–9 and 20 for Plants 1 and 2, respectively. This sort of information is normally identified using a time-series plot, which shows the high values more clearly.

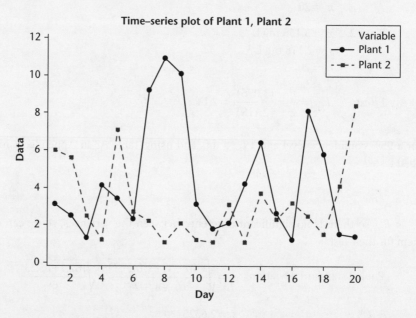

- *to check whether the mean concentration of nitrate in drinking water, measured over the 20 day period from Plant 1, exceeds a proposed new regulatory limit of 10 mg L⁻¹.*

We want to know whether the mean of Plant 1 exceeds the regulatory limit, an example of a one-sided test.

Plant 1: $\bar{x} = 4.255$ mg L^{-1}
$s = 3.096$ mg L^{-1}
$n = 20$
$t_{calc} = \dfrac{\bar{x} - \mu}{s/\sqrt{n}} = -8.30$

In this case we need the one-sided value of t_{crit} for a 95% level of confidence ($p = 0.05$) and ($n - 1$) degrees of freedom, obtained from the t-distribution in Table A1 of Appendix A (note that you must use a different column to the two-sided test);

$$d.f. = 19$$
$$t_{crit} = 1.729$$

In this case $t_{calc} < t_{crit}$, so Plant 1 does not exceed the proposed new regulatory limit.

In addition $t_{calc} < -t_{crit}$ so we can also say that the mean concentration of nitrate is significantly below the limit. However, there are isolated instances on days 8 and 9 when it exceeds the limit, and replicate measurements would have to be made to determine if this is significant.

Sampling error

Learning outcomes

Once you have read through this chapter, worked through the activities, and reviewed the feedback, you should be able to:

1. understand the importance of sampling error;
2. understand how sample size can affect sampling error;
3. calculate the size and number of samples necessary for a given sampling error.

8.1   Introduction

In Chapter 2 we introduced the concept of sampling using a semi-quantitative approach to the subject. Now, having studied the use of basic statistics to evaluate analytical data, it is worth returning to the subject of sampling and using this knowledge in a quantitative way to examine the concept of **sampling error**.

8.2 **What is sampling error?**

In the theory presented by Pierre Gy[1], **correct sampling** is defined as a sampling scenario in which all particles have the same probability of being included in the sample as that in the bulk or source material. Gy's theory addresses seven types of sampling error that can occur during the sampling and handling process. Examples of these types have been presented in some of the case studies and scenarios already discussed in Chapter 2, and while proven techniques for their minimization are known, some sampling errors are difficult to reduce. The seven major types of associated sampling error are:[2]

---

[1] Gy, P.M. 1982. *Sampling of Particulate Materials, Theory and Practice*. 2nd revised ed. Amsterdam: Elsevier; Pitard, F. 1993. *Pierre Gy's Sampling Theory and Sampling Practice: Heterogeneity, Sampling Correctness, and Statistical Process Control*. 2nd ed. Boca Raton, FL: CRC Press.

[2] See http://www.itrcweb.org/ism-1/references/csutpg.pdf

Fundamental error: This error involves the inherent variation in the sample due to its chemical and physical composition. At some level, in solids, liquids, and gases, they may be described as heterogeneous and will involve the size of sample taken. In solids, this includes the particle size distribution of both the analyte of interest and the bulk source. This sampling error may be reduced by decreasing the diameter of the largest particles involved or by increasing the sample mass taken.

Grouping and segregation error: This error arises due to non-random distribution effects and results from **zoning**. This can include the effects from gravity, thermal diffusion, compaction, vibration, weathering, etc. on the bulk source. We saw an example of this in our steel sampling Example 2.1 in Section 2.3.3. It can be minimized by creating a composite analytical sample from many randomly selected increments or by properly homogenizing and splitting the sample.

Long-range heterogeneity error: This is a fluctuating, non-periodic, and non-random effect. It is spatial and may be identified by various continuous sampling regimes. An isolated pollution event into a river could demonstrate this effect. Samples taken at different times can present this, demonstrated as changes in composition. Changes and trends can be followed (and possibly reduced) by taking many increments to form the sample for analysis.

Periodic heterogeneity error: This fluctuation error may be considered a result of cyclic variation; it is temporal or spatial in character and can be minimized by forming composite analytical samples correctly. The frequency of sampling compared with the periodic change (frequency of the cyclic variation) can be critical in this process. The discharge of waste waters from an industrial process to a river, which occurs on a regular and periodic basis, could show this effect.

Increment delimitation error: This error is tied to inappropriate sampling design and possibly the wrong choice of equipment. For example, random sampling of a contaminated field may miss the contaminant and/or the contaminant's highest value.

Increment extraction error: This error occurs when the sampling procedure fails to precisely extract the intended increment; quite often because the sampler being used within that protocol is not fit for purpose. Well-designed sampling equipment and good protocols are crucial.

Preparation error: This error arises due to loss of integrity of the sample. It is often expressed in terms of loss, contamination, and alteration of a sample or sub-sample. Field and laboratory techniques exist to address this problem.

The above seven errors together with what is termed the weighting error, that is, the result of errors in assigning weights to different parts of an unequal composite sample, may actually be considered in terms of even broader types of error.[3] With the exception of the preparation error identified above, most sampling errors are due to the material heterogeneity, itself being divided into two classes:

- constitution heterogeneity, which refers to the fact that all natural materials are heterogeneous, that is, they consist of different types of particles (molecules, ions, grains, etc.) and;

- distribution heterogeneity, where the distribution is heterogeneous if the particles are not randomly distributed in the sampling target (or lot) to be investigated.

Also, the total sampling error can be divided into **correct** sampling errors and **incorrect** sampling errors. The latter, comprising the last three errors in the list of seven shown above

---

[3] Ramsey, M.H. and Ellison, S.L.R. 2007. Measurement uncertainty arising from sampling: a guide to methods and approaches. In: *Eurachem-CITAC Guide UK*.

together with the weighting error result from sampling equipment and procedures that do not follow the rules of sampling correctness, defined in the sampling theory. The first four errors in the above list can actually be modelled and assessed in terms of measureable errors.[3]

One final point should be noted when defining the above errors. The term **variation** is used by Gy and others to describe those processes where they contribute to 'a difference which may be described in terms of an error'. Knowledge of the analyte distribution in the sample in temporal and spatial terms may be an integral part (the reason for or the objective) of the measurement. Hence some errors may be re-appraised in terms dependent upon the scale of the sampling undertaken and framework of the plan.

8.2.1 Magnitude of error

In Chapter 1 we saw that the analytical approach involved a series of steps: to define the problem, then sampling, sample preparation, then measurement, etc. If the reason for the analysis was correctly identified, then all the other steps would contribute to the overall error in the analysis. In Chapter 7 we learned that the mean value of sample analyses, $\bar{x}$, the standard deviation, s, and the variance, s^2, can be used to describe the variability in a set of data. Now, it so happens that the overall variance, which is a measure of the indeterminate or random errors associated with the process, is the sum of the individual variances from each step. When the system under investigation can be defined in terms of a normal distribution we can express the total variance as:

$$s_{tot}^2 = s_{sam}^2 + s_{sp}^2 + s_{meas}^2 \qquad (8.1)$$

where s_{sam}^2 = the variance associated with the sampling step, s_{sp}^2 = the variance associated with the sample preparation step, and s_{meas}^2 = the variance associated with the measurement step.

In many cases, it has been shown that $s_{sam}^2 \gg s_{sp}^2 \geq s_{meas}^2$. So it is not uncommon for the sampling errors to be an order of magnitude greater than the combined sampling preparation and measurement errors. It is instructive to see the effect that changing the standard deviation of the component steps has upon the total variance and vice versa.

So, if $s_{tot}^2 = s_{sam}^2 + (s_{sp}^2 + s_{meas}^2) = 10 + 1$

$s_{tot}^2 = 11$

$s_{tot} = 3.32$

The effect of changing the proportion of the variance in the sampling step is shown in Table 8.1. This demonstrates how the sampling-step error can dominate a system despite the efforts put in to reduce errors associated with the other two steps. In the absence of certain knowledge about the distribution of the analyte, estimates of mean and standard deviation can have even

Table 8.1 Effect of the variance of sampling (s_{sam}^2) compared with combined variance of sample preparation and measurement ($s_{sp}^2 + s_{meas}^2$) on overall precision (s_{tot})

s_{sam}^2	$s_{sp}^2 + s_{meas}^2$	s_{tot}^2	s_{tot}	% improvement to s_{tot} (10 + 1)
10	1	11	3.32	0
5	1	6	2.45	26
10	0.5	10.5	3.24	2.4
5	0.5	5.5	2.34	29.5

greater associated errors and wider limits. Careful attention to the sampling design allows this sampling error to be reduced and ideally, kept to a minimum.

Key points to note with regard to sampling errors are therefore:

1. The sampling step potentially introduces the greatest uncertainty to the analytical process.

2. It is necessary to acquire a representative sample, whose chemical and/or physical characteristics replicate that of the analyte in the bulk material.

3. It is necessary to acquire a homogeneous sample in which the analyte is evenly distributed and is representative of the bulk for the size of sample taken for analysis, which may require some form of sample preparation.

4. In order to achieve 2 and 3, it is necessary to consider how many **sample increments** should be taken, and how much each increment should be.

These laws, associated with probability and distribution, are critical to an understanding of sampling. In order to illustrate what would otherwise be a tedious statistical treatment, we will take two examples to illustrate the point.

EXAMPLE 8.1

Sampling a beach

Consider the example where an analyte is to be determined in sand from a beach, but only one in every 1000 particles of sand has the analyte present.

Now, the probability (p) that a particle containing the analyte is picked out is clearly 1/1000 (or 0.001). By the laws of combining probabilities, the probability that two particles containing the analyte are picked out is $1/1000 \times 1/1000$. So, in general, the probability that n particles chosen at random will all have the analyte will be of the form:

$$p = \left(\frac{1}{1000} \right)^n \tag{8.2}$$

By default, in the 1000 particles, 999 will not contain the analyte, so the probability for n particles not containing the analyte will be given by:

$$q = \left(\frac{999}{1000} \right)^n \tag{8.3}$$

For a simple two-component system, where, by proportion, $p + q = 1$ (e.g. 0.001 + 0.999) and where you either pick a grain with the analyte or not, then the distribution follows a 'binomial' of the form $(p + q)^n$, for n items drawn from the population. The following equations can be used to calculate statistics for particles with and without the analyte:

	With analyte	Without analyte	
Mean	$\bar{x} = np$	$\bar{x} = nq$	(8.4)
			(8.5)
Variance	$s^2 = npq$		(8.6)
Standard deviation	$s = (npq)^{1/2}$		(8.7)
Relative standard deviation as a percentage	$RSD = \dfrac{s}{\bar{x}} \times 100$		(8.8)

So, using our example where $p = 0.001$ and $q = 0.999$, if we pick up 100 000 grains of sand (10^5 particles; and assuming we haven't lost the will to live!) then the variance and standard deviations for the particles with and without the analyte can be calculated and are shown in Table 8.2.

Table 8.2 Sampling statistics for 100 000 particles

	With analyte	Without analyte
$\overline{x}$	100	99 900
s^2	99.9	99.9
s	9.99	9.99(5)
RSD	9.99%	0.01%

Now, if we take 100 000 grains of sand it is worth considering the following before proceeding:

1. If the sand grains are assumed to be spherical and 1 mm in diameter and have a density of 2.65 g cm^{-3} this equates to a sample mass of 138.8 g, of which only 0.14 g (100 particles) contain the analyte.

2. The variance and standard deviation in terms of numbers of grains are the same for both, that is, ±10 grains or ~ ±0.014 g of sand, which is the error associated with drawing the analyte of interest from 10^5 particles of sand, of which only 100 actually contain it.

3. In relative terms the grains with the analyte have an RSD of ~10%, whereas the grains without the analyte have an RSD of 0.01%.

4. An analytical method that gives an RSD of 10% would not normally be considered under full analytical control.

5. A confidence interval based upon ±3s would present a 60% range of values (100 ± 30 grains of sand), a poor outcome for a sample which weighs ~140 g!

Clearly, the number of particles and the sampling statistics are interlinked. So how might we change the system in our favour? There are two basic ways to improve the sampling statistics:

- by increasing the number of particles in the sample;
- by producing more, smaller particles by grinding the original sample.

The advantage of grinding is that you would be able to increase the number of particles but reduce the sample weight to a value which is analytically acceptable, while keeping the variance under control as well. So, using the same example, if we grind our original bulk material so that the grains are now 0.1 mm diameter, then 100 000 grains would now weigh only 0.1388 g instead of 138.8 g; a 1000-fold decrease. But the important point is that the sampling statistics will be the same. Laboratory sub-samples (sometimes known as test samples) of particulate material typically weigh between 0.25 and 10 g. So, if we grind the sand sample down to a particle size of 0.1 mm diameter and take 0.971(3) g (i.e. 7×0.1388 g) this sub-sample would contain 700 000 particles. If we repeat our statistical analysis from earlier it results in the data shown in Table 8.3.

Table 8.3 Sampling statistics for different sample masses and particle sizes

No. of particles sampled (n)	Particle size/mm	Mass of sample/g	No. of particles containing analyte	s^2	s	RSD/%
100 000	1.0	138.8	100	99.9	9.99	9.99
100 000	0.1	0.1388	100	99.9	9.99	9.99
700 000	0.1	0.9713	700	699.3	26.45	3.78

As can be seen, the RSD is now 3.78%, which is much more acceptable than ~10%. So, by increasing the number of particles sampled by a factor of 7 we have reduced the RSD by a factor of $\sqrt{7} = \sim2.65$ times.

By this example we can see that the questions 'How much and how many samples?' depend on what is an acceptable level of accuracy and precision. In other words, the more samples you take **correctly**, the greater the confidence that can be provided.

EXAMPLE 8.2

Sampling ore

Sampling from mining ore presents a slightly more complex case of the previous example, because now the bulk material, while being particulate in nature, will also vary in size, morphology and composition to a greater degree; in other words it is a much more **heterogeneous** material. So, the particle size distribution of the sample components, the distribution of the analyte within those component particles, the density of each component type, and the required or accepted level of sampling variance must all be considered to provide us with an estimate of the required **gross sample** size.[4]

In this case, the variance may be calculated using:

$$s^2 = \left[\frac{1}{m_s} - \frac{1}{m_L} \right] K d_n^3 \tag{8.9}$$

where m_s = mass of sample in g, m_L = mass of the material from which m_s is taken in g, K = a constant for a given particulate material, and d_n is the nominal size of particulate fragments in the sample. In most practical cases $m_s <<< m_L$, often by a factor of 500–1000, such that the equation simplifies to:

$$s^2 = \frac{K d_n^3}{m_s} \tag{8.10}$$

where $K = f \times g \times c \times l$.

f = shape factor, which is between 0.1 and 1, and depends upon the morphology of the particulate material (needles = 0.1 and a cube = 1); it is often taken to be ~0.5;

g = particle size range factor (or granulometric factor), which is between 0 and 1, where low values for g denote a large range of particle sizes and high values denote a small range and identical particles = 1; it is often taken to be close to 0.25 for generally un-crushed samples with a broad size range;

[4] See the references for P. Gy, for D. Francois-Bongarcon, and for F. F. Pitard, all in the paper by Minnitt, R.C.A., Rice, P.M., and Spangenberg. C. 2007. Part 1: Understanding the components of the fundamental sampling error: a key to good sampling practice. *The Journal of The Southern African Institute of Mining and Metallurgy*, 107(8), pp.505–511

c = composition factor, which depends upon the mineralogical composition but can be approximated by dividing the density (g cm^{-3}) of the material containing the analyte by the concentration in mg kg^{-1} (e.g. if a gold nugget (density 19.3 g cm^{-3}) is in a material at a concentration of 1 mg kg^{-1} (1 in 10^6 g) then $c = 19.3/(1 \times 10^{-6})$ = 19.3×10^6 g cm^{-3});

$l = \left[d_l / d_n \right]^b$, the liberation factor, which has values between 0 (no liberation) and 1 (complete liberation). The liberation size, d_l, is the nominal size to which fragments must be crushed in order to fully liberate the analyte from the matrix. The exponent b has values between 0 and 3 depending on type of solid, for example, for gold-containing ores b is ~1.5 and may be used as a general starting point.

In most practical cases, the particle size that allows 95% of the material to pass is considered a suitable value for d_n. Other approximations can also be made to arrive at the data given in Table 8.4. If nothing else, this approximation demonstrates just how large the sample weight can be for a given size of bulk material.

Table 8.4 Minimum sample size relative to the bulk material (approx. guide)

Material bulk weight (tonnes)	Sample weight (kg)
up to 0.5	2.5
0.5–1.0	4
1.0–2.0	6
2.0–5.0	10
5–10	15
10–25	25
25–50	40
50–100	60

8.3 Sampling for a given level of precision

Rather than calculate the precision inherent in any given sampling regime, it is considerably more useful to design a sampling plan that results in a defined error which is acceptable. In order to do this, two factors are under our control, the size and number of samples.

8.3.1 How much sample?

Returning to our simplest case, where an analyte is distributed in grains of sand, we can calculate how much sample to take for a required level of precision. Let us assume that we require a level of 10% RSD in the measurement and that the density of all particles is the same throughout. If we express the RSD as a fraction (s_r) rather than a percentage we can say that if RSD = 10%, $s_r = 0.1$. So, for a single analyte,

$$s_r = \frac{s}{np} \tag{8.11}$$

where s_r = RSD expressed as a fraction, p = the probability of picking a grain containing the analyte, s = standard deviation, and n = the total number of particles sampled.

In our sand example, $p = 1/1000$ (i.e. 0.001) and 100% of the analyte is contained in every 1 in a 1000 grains and 0% in the other 999.

In a simplified population sampling system with a normal distribution, the standard deviation is given by:

$$s = (npq)^{\frac{1}{2}} \tag{8.12}$$

where n = total number of particles sampled, p = fractional probability of sampling the analyte particle, and q = fractional probability of sampling a particle not containing the analyte. Substituting for $q = 1 - p$ we obtain

$$s = \left[np(1-p) \right]^{\frac{1}{2}} \tag{8.13}$$

Substituting equation (8.13) into equation (8.11), simplifying, and rearranging:

$$s_r^{\,2} = \frac{1-p}{np} \tag{8.14}$$

$$s_r = \left(\frac{1-p}{np} \right)^{\frac{1}{2}} \tag{8.15}$$

$$n = \frac{1-p}{s_r^{\,2}\,p} \tag{8.16}$$

Hence, the number of particles (n) required to give us the level of precision can be estimated, based on probability, to be

$$n = \frac{(1-0.001)}{0.001 \times 0.1^2}$$

$$n = 99\,900 \text{ particles}$$

Since each 1 mm grain of sand weighs 0.00138(8) g then 99 900 grains will weigh 138.7 g; which is exactly what we saw in the 'beach case' before. So, simplifying our equation and knowing that:

$$W = n \rho_m V \tag{8.17}$$

where W = mass of material, n = number of particles, ρ_m = density of material, and V = volume of assumed spherical particles.

Then the full equation becomes

$$W = \frac{(1-p)}{p.s_r^2} \rho_m V \tag{8.18}$$

or

$$W = \frac{(1-p)}{p.s_r^2} \rho_m \frac{\pi d^3}{6} \tag{8.19}$$

This equation gives the mass (or weight) of sample needed to provide the relative variance or standard deviation for a given material of density ρ_m and mean particle diameter d.

ACTIVITY 8.1 To help illustrate the problem of taking a representative sample (II)

[5]Using the tabulated data obtained from Activity 2.1, compare the sample standard deviation (s) and RSD (s_r) values with the theory. Theoretically for a two-component system like this the RSD is given by equations (8.14) and (8.15) to be:

$$s_r^2 = \frac{1-p}{np}$$

$$s_r = \left(\frac{1-p}{np}\right)^{1/2}$$

where p is the fractional concentration and n is the number of beads sampled.

1. For each of your three sample sizes, calculate the theoretical RSD and compare it to your calculated RSD.

2. Use the mean for p, and calculate the average number of beads sampled for this beaker size for n. Does your data agree with the overall trend predicted for the RSD as a function of sample size?

3. Now look at the individual and combined groups' data and compare. Is the predicted trend clearer when a larger number of measurements are included?

4. Perform a t-test at the 95% confidence level to compare your sampling mean with the target fractional concentration of 0.10 ± 0.01 (i.e. the sample standard deviation = 0.01). Now discuss the answer to the question 'is the concentration of chlorpheniramine maleate, represented by the "active" (e.g. red) beads, outside the specification?'

8.3.2 How many samples?

How many samples should be taken to give the required variance? Assuming that we have collected enough sample, in the correct way, the indeterminate errors will still limit the confidence in our sample's value. Therefore, we may consider the following iterative process.

Using the equation for the t-test discussed in Section 7.4.1 we can derive an alternative form for this expression:

$$\mu - \bar{x} = \frac{ts}{\sqrt{n}} \tag{8.20}$$

If we express this in relative terms, as in the expression for the relative standard deviation given in equation (7.7), we arrive at

$$\left(\mu - \bar{x}\right)_r = \frac{ts_r}{\sqrt{n}} \tag{8.21}$$

---

[5] Based upon the work of Vitt, J.E. and Engstrom, R.C. 1999. Effect of sample size on sampling error: an experiment for introductory analytical chemistry. *Journal of Chemical Education*, 76(1), pp.99–100 and also F. Cheng at http://www.webpages.uidaho.edu/ifcheng/Chem%20253/labs/Chem%20254.htm (Experiment 2 2015-02-13.docx).

We can also set the condition that the maximum desired value for the RSD is equivalent to this difference between the mean and the true value, so we can say that

$$s_{rd} = (\mu - \bar{x})_r \qquad (8.22)$$

$$s_{rd} = \frac{ts_r}{\sqrt{n}} \qquad (8.23)$$

By rearrangement we obtain the following expression:

$$\frac{n}{t^2} = \frac{s_r^2}{s_{rd}^2} \qquad (8.24)$$

Using our previous case of taking 100 000 sand grains from the beach, each 0.1 mm diameter (Table 8.3) and where the analyte is contained in only 100 grains, we saw that 0.138 g gave an RSD of 9.99% (i.e. $s_r = 0.1$) for the analyte. So, to achieve an RSD of 3.78% ($s_{rd} = 0.0378$), how many 0.138 g samples would we need to take, at a confidence level of 95% ($p = 0.05$)? On the face of it, we simply need to substitute the relevant values into equation (8.24) and solve for n. However, because the value of t also depends on n, the solution must be arrived at as follows:

$$\frac{n}{t^2} = \frac{0.1^2}{0.0378^2} = 6.9987$$

The relationship between t and the number of degrees of freedom (d.f.) for $p = 0.05$ is given in Table A1 in Appendix A. In this case, d.f. $= n - 1$ so $n =$ d.f. $+ 1$, and we can use this data to plot d.f. versus (d.f. $+ 1)/t^2$, as shown in Figure 8.1.

Then: $\qquad \dfrac{\text{d.f.} + 1}{t^2} = 6.9987$

Interpolating: d.f. $= 28$

so: $n - 1 = 28$

and: $n = 29$

Thus, we need to take ~29 samples, each of 0.138 g, in order to achieve ±3.78% RSD with 95% confidence. This RSD of 3.78% is, in theory, also achievable by taking just one larger sample of 0.971(3) g if the number of particles associated with the analyte is increased by a factor of ~7 in proportion to the total number of particles in the sample; a result of the analyte's distribution.

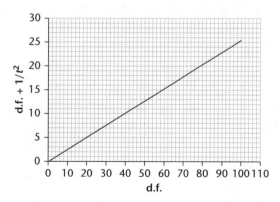

Figure 8.1 Plot of d.f. versus (d.f. $+ 1)/t^2$

Of course, if we require a greater level of confidence, such as 99.8%, in achieving an RSD of ±3.78% then we would require 71 samples, each of 0.138 g or ~3 samples of 0.971(3) g.

ACTIVITY 8.2 To help illustrate the problem of taking a representative sample (III)

Using the data you obtained from Activity 2.1 and considering the worked values from Activity 8.1 then, quantitatively, it can be shown that the (relative standard deviation)^2 × the number of beads in a sample, n, should be a constant for a specific analyte, (called the 'sampling constant'). Estimate this constant by first calculating the constant from each sampling size experiment (10 mL, 20 mL, and 50 mL) and then average the three values to obtain an overall estimated experimental 'sampling constant' value. Then use this experimental constant to predict the sample size, n, which should be used to obtain a %RSD of 1 and 0.1; that is, use the same equation and solve for the number of beads with a %RSD now set to their fractional values of 0.01 and 0.001.

Feedback on activities

 FEEDBACK ON ACTIVITY 8.1

To help illustrate the problem of taking a representative sample (II)

Take the example shown in Activity 2.1 as the basis for our scenario, that is,

- Take a large bag containing around 1000, ~5 mm sized inert coloured beads which will be used to represent the different components in a sample.

- The beads should all have the same size and mass and therefore their per cent by number, per cent by weight, and per cent by volume are all identical.

- Consider a system with just two colours where one colour represents the main sample matrix (e.g. white beads), which, for example, can be a harmless formulation used to make, say, antihistamine tablets, while the other colour (e.g. red beads) represents, say, the pharmacologically active ingredient, chlorpheniramine maleate, which must be present at a fractional concentration of 0.10 ± 0.01, that is, one 'active' (e.g. red) bead for every 10 beads in the bag (10%; 9 white beads).

- Use three different sized volume capacity plastic beakers to sample the mixed set of beads in the bag, 10 mL, 20 mL, and 50 mL.

Therefore, from our theory that for a two-component (binomial) system and equations (8.12) and (8.15):

$$s = \left(npq\right)^{\frac{1}{2}}$$

$$s_r = \left(\frac{1-p}{np}\right)^{\frac{1}{2}}$$

if p is the fractional content of the active ingredient such that

$$p = \frac{n_a}{n_a + n_m}$$

where n_a = number of 'active' beads and n_m = number of matrix beads, and n is the number of beads actually sampled, then if the *whole* bag was sampled, the theoretical standard deviation (SD) for the active component is

$$s = (1000 \times 0.1 \times 0.9)^{1/2} = 9.49$$

and the RSD for the active component, as a fraction, is

$$s_r = \left(\frac{1-0.1}{1000 \times 0.1} \right)^{1/2} = 0.0949$$

or 9.49%.

For our practical experiment, as the number of beads sampled increases from using the 10 mL to the 20 mL and finally the 50 mL plastic beaker, there is a theoretical number of beads acquired for each under 'ideal conditions'. This can be calculated based upon packing of the beads within the container and gives rise to differences in the sampling number, n. For random packing, the volume occupied depends upon the packing density such as:

Type of packing	Example	Packing density*
Very loose random packing	Spheres slowly settled	0.56
Loose random packing	Dropped into bed or packed by hand	0.59–0.60
Poured random packing	Spheres poured into bed	0.609–0.625
Close random packing	The bed is shaken/vibrated	0.625–0.641
Densest regular packing	Close packed (co-ord. no. 12)	0.7405

*Dullien, F.A.L. 1992. *Porous Media: Fluid Transport and Pore Structure*. 2nd ed. Academic Press.

For general practical purposes, the random packing density is often taken to be 0.64 (i.e. 64% packed and 36% void volume). Based upon 5 mm beads, each with a volume of 0.06545 mL, then, as an example, the theoretical number of beads in the 10 mL beaker could be:

$$[10 \text{ mL} \times 0.64]/0.06545 = {\sim}97\text{–}98 \text{ beads}$$

Using the equations shown in the previous section where we sampled the entire bag of beads, the calculated theoretical values for SD and RSD can be obtained for the 10 mL, 20 mL, and 50 mL sampling volumes by inserting the relevant numbers. These can then be compared with the experimental SD and RSD values obtained in the class.

? **FEEDBACK ON ACTIVITY 8.2**

To help illustrate the problem of taking a representative sample (III)

If we consider the theory upon which this sampling experiment is based, then rearranging equation 8.16, it can be seen that the product of the total number of particles sampled, n, and the square of the relative standard deviation (RSD^2), shown as s_r^2, will be a constant;

$$n \times s_r^2 = \text{constant} = \frac{[1-p]}{p}$$

This is because our simplified theory states that the fractional probability, p, of drawing our analyte from a two component system is a constant. We have called this the 'sampling constant'.

In our sand sampling example where we actually used equation 8.16, the fractional probability of 0.001 for our analyte (1 in every 1000 sand particles) gives rise to a 'sampling constant' of 999 based upon the required level of 10% RSD (fractional RSD of 0.1).

The sample size for our beads, n, can therefore be calculated using the sampling constant from the experiment by setting the %RSD as its fractional value (0.1% = 0.001 and 1% = 0.01; for the two separate cases in the activity) and solving in the above equation.

Now, it can be shown from Activity 8.1 that the three different sample sizes chosen (10 mL, 20 mL, and 50 mL) result in three very different numbers of beads (particles) being acquired. However, each beaker volume produces a near constant number of beads sampled, with a little variation. As the beads are identical in size (i.e. diameter, and being spherical also in terms of volume) and with them being made of the same material and having the same density, the mass of each particle is also a constant. In this simple binary system (particles are red or white in our activity) each particle has the same size and mass. Therefore, their % by number, % by volume, and % by weight values are all identical. As a result, it is possible to simplify those descriptive equations that in the real world are far more complex. Hence, we can state, for a start, that the total mass sampled in a beaker is proportional to the number of beads (particles) in each case.

Considering equation 8.19 and how it was derived from equation 8.16, we can see that the product of the sample mass, W, and the (relative standard deviation)^2, s_r^2, also gives us a constant, because as before, all the other components in the equation (8.19) are themselves constants. We can therefore state, in this idealised system, that:

$$W \times s_r^2 = \text{constant} = K_s$$

This constant Ks is called the *Ingamells sampling constant*[6]. Its derivation, its use and its equivalent variants can be found in a number of publications since its introduction.

In the real world, a solid sample may present a multi-component system with a range of composition and density and particle size. Using equation 8.19, it may also be possible to estimate the particle size (diameter) that a sample's components should be reduced to (by grinding and sieving), in order to provide an accepted level of sampling precision as well as the overall sample mass required for that same accepted level of sampling precision.

---

[6] Ingamells, C.O. 1974. New approaches to geochemical analysis and sampling. *Talanta*, 21, pp.141–155 and also Ingamells, C. O. and Switzer, P. 1973. A proposed sampling constant for use in geochemical analysis. *Talanta*, 20, 547–568.

Method validation and quality assurance

9

Learning outcomes

Once you have read through this chapter, worked through the activities and problems, and reviewed the feedback, you should be able to:

1. explain the concept of method validation and the role of performance characteristics;

2. calculate the performance characteristics for an analytical method;

3. explain the concepts of quality management and quality assurance with respect to analytical measurements;

4. plot Shewhart and moving range quality control charts;

5. interpret a quality control chart used for quality assurance of an analytical method.

9.1 Introduction

Before you can use an analytical method to produce reliable results, it must be shown to be fit for the purpose for which it is intended. More specifically, in the context of this chapter, **fitness for purpose** means that:

> Analytical measurements should be made using methods and equipment which have been tested to ensure they are fit for purpose.[1]

The process required to demonstrate this is called **method validation**. An important part of method validation is the generation of data used to determine if the **performance characteristics** of the method are satisfactory.

---

[1] Magnusson, B. and Örnemark, U. (eds.) 2014. *The Fitness for Purpose of Analytical Methods: A Laboratory Guide to Method Validation and Related Topics*. 2nd ed. Available at: www.eurachem.org.

9.2 Performance characteristics

Table 9.1 lists the performance characteristics that need to be checked during method validation.

The exact process of method validation will depend on the analytical method, its scope, the analyte to be determined and the nature of the sample. In order to help us understand this in its wider context let us consider the determination of the compounds shown in Figure 9.1. All of these compounds contain phosphorus but they are all very different compounds, so are therefore different **analytes** which will require different **analytical methods**. We will use these as examples to illustrate the various steps in the process of method validation.

Table 9.1 Performance characteristics

Performance characteristic	Procedure that could be used
Limit of detection	Determine sensitivity and standard deviation of replicates of a calibration standard close to zero concentration
Limit of quantitation	Determine sensitivity and standard deviation of replicates of zero standard
Bias estimate	Analyse a certified reference material (CRM) or spiked sample and perform a statistical test
Linear range	Perform a calibration over the method range
Selectivity	Analyse a CRM or spiked sample and perform a statistical test Check selectivity of the instrumental method
Precision	
Repeatability	Repeat analysis of a CRM or homogeneous sample
Reproducibility	Several laboratories perform repeat analyses of a CRM or homogeneous sample
Ruggedness	Identify components of the method that require special control

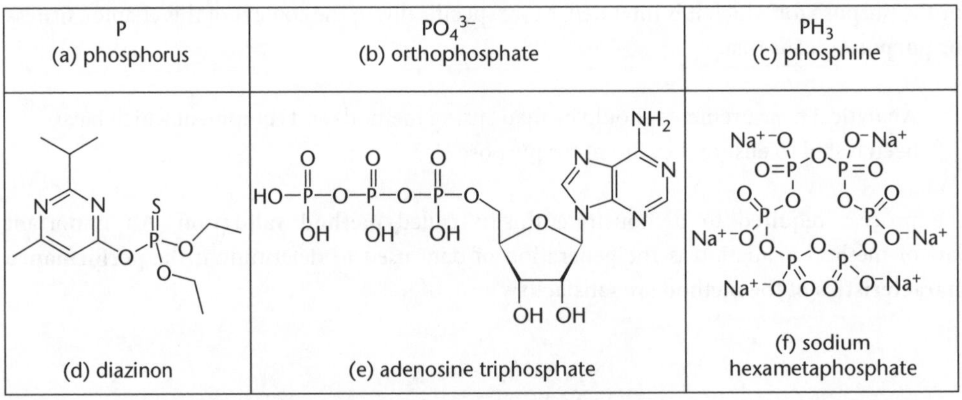

Figure 9.1 Phosphorus-containing compounds

ACTIVITY 9.1   Which method?

For each of the compounds in Figure 9.1:

- Identify the analyte.
- Assign the analyte to the correct instrumental measurement method from a choice of: liquid chromatography; UV-vis spectroscopy; bioluminescent enzyme assay; GC-MS; ICP-AES; FAAS.

Compound	Analyte	Method
(a)		
(b)		
(c)		
(d)		
(e)		
(f)		

9.2.1 Limit of detection

Sensitivity is the change in instrument response (dy) per unit concentration (or quantity) of analyte (dx) and is given by the slope of the calibration curve ($b = dy/dx$), as shown in Figure 9.2. Sensitivity is often confused with the **limit of detection (LOD)**, the power of detection of any method of analysis. However, they are not the same thing: the LOD takes account of both sensitivity and random fluctuations in the signal. The LOD is derived from the smallest measure of the signal which can be accepted with confidence as genuine and can be distinguished from the noise so

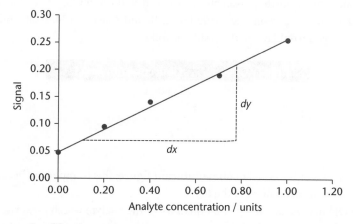

Figure 9.2 A generic calibration curve

it is not suspected to be only an accidentally high value. The value at the 99.7% confidence level (the so-called 3s level) is usually taken to indicate this:

$$y_{LOD} = \bar{y}_0 + 3s_0 \qquad (9.1)$$

where $\bar{y}_0$ is the mean and s_0 is the **estimate** of the **sample standard deviation** (σ_{n-1}) of ten replicate measurements of a standard at or near zero concentration.

Remember that, from Chapter 7, if enough replicate measurements are made then 68.3% of the readings lie within ±1s (one standard deviation) of the mean, 95.4% within ±2s, and 99.7% within ±3s. So, when the LOD is calculated using equation (9.1) there is only a 0.15% chance that a reading greater than $\bar{y}_0 + 3s$ is due to an unusually high blank reading. In this case we can say that the reading is greater than the LOD and it is a valid result.

In practice, it is much more useful if we express the LOD in concentration units (x_{LOD}) which we can calculate using the slope (b) of the calibration thus:

$$x_{LOD} = \frac{3s_0}{b} \qquad (9.2)$$

which is normally just referred to as the LOD.

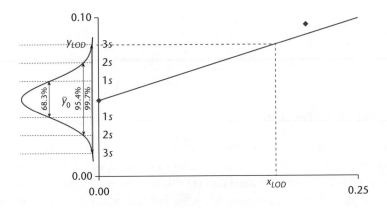

It is important to note that if the measurement reading for a particular sample gives a result that is below the LOD we would give the result as <LOD. For example, when using a method for the determination of orthophosphate in water, if you established that the LOD was 0.1 mg L^{-1} and the calculated results for samples A, B, and C were 0.34, 0.08, and 5.5 mg L^{-1}, respectively, then you would quote the results as follows:

Sample	Concentration/mg L^{-1}
A	0.34
B	<0.1
C	5.5

Note that the LOD is only quoted to **one significant figure**; this is important because it reflects the precision of the measurement at the LOD.

Thus, the LOD may be defined 'as that quantity of the analyte which gives rise to a reading equal to three times the standard deviation of a series of at least ten determinations of the standard at or near zero concentration'. So, we can see that the LOD takes into account both the sensitivity and the stability of the instrument used for the measurement, and it can vary from day to day and between instruments of the same type.

When faced with a real analysis of real samples, the **limit of quantitation (LOQ)** is the lowest concentration of an analyte that can be determined with acceptable precision. This is often a matter of judgement but a good rule of thumb is to take the LOQ as being $10s_0$ ($10\times$ the standard deviation of the standard at or near zero concentration). So:

$$x_{LOQ} = \frac{10s_0}{b} \tag{9.3}$$

which is normally just referred to as the LOQ.

ACTIVITY 9.2   Calculating the limit of detection

Orthophosphate must be determined in river water because runoff from agricultural land which has been treated with phosphate-containing fertilizers can lead to environmental problems such as eutrophication. **UV-vis spectroscopy** can be used with the 'molybdenum blue' method (in which orthophosphate reacts with molybdate in an acid medium to form a phosphomolybdate complex which is then reduced by ascorbic acid, resulting in an intense blue colour) to directly determine orthophosphate in water samples. You have performed a calibration of the instrument and obtained the data given in Figure 9.3 and Table 9.2.

The equation of the calibration is $y = 0.21x + 0.048$ (Figure 9.3), so the slope is given by $b = 0.21$ absorbance mg^{-1} L (absorbance per $mg\ L^{-1}$).

- Calculate the sample standard deviation, s_0.
- Calculate the LOD using equation (9.2).
- Calculate the LOQ using equation (9.3).

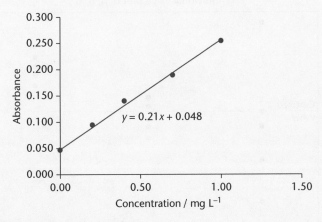

Figure 9.3 Calibration for orthophosphate

Table 9.2 Results of ten replicate measurements of the zero orthophosphate standard

Replicate	Absorbance
1	0.044
2	0.056
3	0.043
4	0.052
5	0.051
6	0.049
7	0.044
8	0.050
9	0.048
10	0.040

9.2.2 Bias

The term accuracy is often used in analytical science to mean **trueness.** However, accuracy involves elements of both **bias** (trueness) and **precision.** The trueness of a measurement result is defined as a consistent difference between the measured value and the true value; if a method consistently produces high or low results it is said to be biased and there is a **systematic error.** Precision is defined as the closeness of agreement between independent repeated measurements and is a measure of **random errors.** The **standard deviation** can be used as a useful measurement of precision.

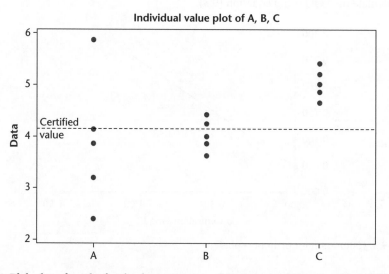

Figure 9.4 Blob plot of results for the determination of diazinon in CRM851 performed by three separate analysts, A, B, and C

For example, look at the results given in Figure 9.4 for the determination of diazinon in the certified reference material CRM851-050 organophosphorus pesticides in soil,[2] performed by three separate analysts, A, B, and C. Each of the analysts has performed five repeat analyses and averaged the results. The certified value for diazinon is 4.14 mg kg^{-1} and is shown as a horizontal dashed line.

Q: What conclusions can you draw about the accuracy and precision of the results for the three different analysts?

A: Analyst A: Unbiased and imprecise—the average result is close to the certified value but it is imprecise because of the large spread.

Analyst B: Unbiased and precise—the average result is close to the certified value and the spread is small.

Analyst C: Biased and precise—the average result is not close to the certified value but the spread is small.

Hence, one way of estimating bias is to analyse a **certified reference material (CRM)** in parallel with your samples to check that the method is giving accurate results. If possible a CRM should be chosen which has a matrix similar to the sample and contains the analyte at a similar concentration.

ACTIVITY 9.3 Estimating bias using a CRM

As we saw in Chapter 7, in order to make an objective decision about the accuracy we need to perform a statistical test, the **t-test**, which takes random errors into account. In the example given above for the determination of diazinon in CRM851-050 (certified concentration 4.14 mg kg^{-1}) the experimental data obtained by the three analysts is given in Table 9.3.

Table 9.3 Results for the determination of diazinon in CRM851-050

Analyst	Measured mean/ mg kg^{-1}	s	n
A	3.89	1.29	5
B	4.03	0.31	5
C	5.04	0.29	5

Calculate whether their measured mean values differ significantly from the certified value at the 95% confidence level. For this comparison of an experimental mean with a known value you will need to use Equation 7.10 in Chapter 7.

9.2.3 Recovery factor

In analytical methods, bias can sometimes be estimated using a **recovery factor**:

$$R = \frac{\bar{x}}{\mu} \tag{9.4}$$

where R = recovery factor, $\bar{x}$ = mean concentration, and μ = true concentration.

---

[2] LGC Ltd. 2017. *LGC Standards*. Available at: https://www.lgcstandards.com/GB/en/

A recovery factor can be determined either by using a CRM or a **spiked sample**. For example, consider the determination of diazinon in the soil sample described in Activities 9.1 and 9.3. The method involves an **extraction** step where the analyte must be extracted from the soil sample using solvent extraction (e.g. the Soxhlet method). You can calculate a recovery factor in two ways:

(a) **Using a CRM**

The certified value for diazinon is 4.14 mg kg^{-1} in CRM851-050 and the measured concentration obtained by Analyst A was 3.89 mg kg^{-1}.

$\bar{x} = 3.89$ mg kg^{-1}
$\mu = 4.14$ mg kg^{-1}

Using equation (9.4):

$$R = \frac{\bar{x}}{\mu} = \frac{3.89}{4.14}$$
$$R = 0.940 \quad or \quad 94\%$$

So we can say that the recovery factor is 0.94, or that the measured value is 94% of the true value.

(b) **Using a spiked sample**

In order to calculate recovery using a spiked sample, one of the samples (A) that is to be analysed would be split in two, then one of these sub-samples (A$_1$) is spiked with a known amount of the analyte (s) while the other (A$_2$) remains unspiked. The concentration of analyte would then be determined in each of the spiked and unspiked samples to yield the data shown in Table 9.4.

The difference between sample A$_1$ and A$_2$ is the measured concentration of the spike, so we can determine the spike recovery factor (R_s) as follows:

$$R_s = \frac{\bar{x}_{A_1} - \bar{x}_{A_2}}{s}$$ (9.5)

where s = concentration equivalent of analyte added to sample

For example, analysis of a spiked sample yielded the following information:

Spiked sample A$_1$: $s = 6.65$ mg kg^{-1}

$\bar{x}_{A_1} = 11.1$ mg kg^{-1}

Unspiked sample A$_2$: $\bar{x}_{A_2} = 5.40$ mg kg^{-1}

Table 9.4 Data matrix for a spiking experiment

Sub-sample	True	Spike	Measured
	Concentration of analyte in sample		
A$_1$	unknown	s	$\bar{x}_{A_1}$
A$_2$	unknown	–	$\bar{x}_{A_2}$

Using equation (9.5):

$$R = \frac{11.1 - 5.40}{6.65} = \frac{5.70}{6.65}$$

$$R = 0.857$$

In this case the recovery factor is 0.857 and the measured concentration of the spike is 85.7% of the actual spike concentration. This method is useful when an appropriate CRM is unavailable or as a method for determining recovery on a routine basis where the supply of CRM is limited.

9.2.4 Linear range

The **linear range** of a method is the range over which the response is proportional to the analyte concentration. In practice, the range of the method covers the lower and upper limits of concentration of the analyte which can be determined with accuracy. This is often defined by the **linearity** of the calibration curve such that the LOQ defines the lower end and the highest calibration standard on the linear part of the calibration defines the upper end, termed the lower and upper limits of quantitation, respectively, or LLOQ and ULOQ.

Q: Using the example given in Activity 9.2 define the LLOQ and ULOQ for the method.

A: LLOQ = 0.2 mg L^{-1}

ULOQ = 1.00 mg L^{-1}

9.2.5 Precision

There are several measures of precision. The simplest is the precision obtained by making replicate measurements on the same sample using the same instrument over a short period of time. However, this only gives us some idea of the instrumental precision and does not tell us much about the precision of the whole method. A more useful measure is to determine the **repeatability** of a method by making successive determinations on separate replicate portions of the sample, by the same analyst, using the same apparatus, over a short time period.

Sometimes you may wish to know how the method performs in the hands of another laboratory; then you would need to determine the **reproducibility** of the method, where analyses are performed by different analysts, using different sets of equipment, over a long time period.

ACTIVITY 9.4 Estimating repeatability and reproducibility

Repeatability

The repeatability of a method for the determination of hexametaphosphate in sausage meat is evaluated by making 10 replicate analyses using ion **chromatography** with a conductivity detector to give the following results:

462	455	491	451	449	473	471	468	466	458	μg g^{-1}

- Calculate the mean, standard deviation, and relative standard deviation.

→
Reproducibility
In an evaluation of the same method for the determination of hexametaphosphate in sausage meat, a sample of sausage was subdivided and sent to 10 different laboratories for analysis. Each lab analysed the sample five times and reported the mean results as follows:

Lab	1	2	3	4	5	6	7	8	9	10
Mean result/$\mu g\ g^{-1}$	452	465	451	491	449	437	417	386	411	401

- Calculate the median, interquartile range, and upper and lower whiskers of a box plot. Identify any possible outliers.

9.2.6 Selectivity

The **selectivity** of an analytical method is the extent to which other substances affect the determination of the analyte. Methods can have varying degrees of selectivity. Some methods, which are extremely selective for a particular analyte, are often called **specific** methods.

EXAMPLE 9.1

Selectivity in spectroscopy
In the determination of orthophosphate in waters using the molybdenum blue method, arsenate (AsO_4^{3-}) produces a colour similar to orthophosphate and might cause a positive interference, though arsenate concentrations up to 100 $\mu g\ L^{-1}$ do not interfere and levels are likely to be well below this in most natural waters.

EXAMPLE 9.2

Selectivity in ICP-AES
To enable reuse of platinum in catalytic converters from scrapped automobiles, the latter must be analysed to determine how much precious metal they have in them when they are recycled. Platinum is determined by ICP-AES after a fire-assay sample preparation which solubilizes the precious metal and the aluminosilicate support material, so there are high concentrations of other elements present. Therefore, it is important that the concentration determined by the analytical method is only due to Pt and is not affected by other components of the matrix. The atomic emission spectrum for Pt in one of the samples is shown in Figure 9.5, and you can see that there is an interference due to Mg in the sample matrix because the high concentration of this element in the sample (from the catalyst support material) gives rise to a small atomic emission signal on the shoulder of the main Pt line at 214.423 nm. So in this case the method is not entirely selective for Pt!

Q: How could you improve the selectivity of the ICP-AES method for Pt?

A: You could improve the selectivity of the method by: separating the Pt from the sample matrix prior to the ICP-AES measurement; increasing the resolution of the detection system to separate the Pt and Mg lines in the spectrum (albeit at the expense of sensitivity); or by using an alternative Pt wavelength which does not suffer from an Mg interference.

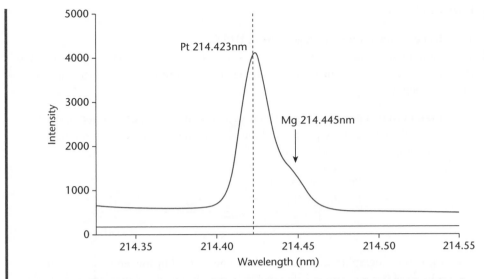

Figure 9.5 ICP-AES spectrum for the determination of Pt in an automotive catalyst sample

ACTIVITY 9.5 Selectivity

P (a) phosphorus	PO₄³⁻ (b) orthophosphate	PH₃ (c) phosphine
(d) diazinon	(e) adenosine triphosphate	(f) sodium hexametaphosphate

Using the feedback from Activity 9.1, describe how the methods are selective for the respective analytes shown above.

For method validation it is necessary to **demonstrate** that a method is selective by analysing a matrix CRM or spiked sample to check the accuracy. Also, you could check for interferences during method validation by analysing spiked samples which contain the analyte and matrix components over the range of concentrations likely to be encountered, and take measures to eliminate any interferences observed (see Chapter 5).

9.2.7 Ruggedness

An analytical method which is considered to be **rugged** is one which performs with the same accuracy and precision despite variations in sample type or size, over time, in different locations, using different instruments, or when some of the method conditions change slightly, for example, temperature or pH.

EXAMPLE 9.3

The effect of temperature on ruggedness in HPLC

The determination of hexametaphosphate in foods uses ion chromatography. The method is prone to a number of variables which must be carefully controlled if it is to be considered rugged, as follows:

- It is known that polyphosphates will hydrolyse at high temperature and low pH, so samples need to be kept cool and the pH should be adjusted with sodium hydroxide.

- The chromatographic separation itself can be affected by the temperature of the column so it is recommended to maintain the column at a constant temperature higher than ambient, for example, 30°C. In fact, changing the temperature in a controlled way can also be used to speed up a separation, much like changing the mobile phase composition.

EXAMPLE 9.4

The effect of pH on ruggedness in HPLC

The interaction of analytes and sample with the stationary and mobile phases is influenced by polarity and pH. Consider the functional groups on the compounds listed below and the reversed phase chromatograms given in Figures 9.7–9.9:

The ionizable compounds have carboxylate groups which will be more or less protonated depending on pH and the pK_a of the compound (Figure 9.6).

In reversed-phase liquid chromatography the stationary phase is usually made of **non-polar** alkyl chains bonded to silica particles, and the mobile phase is a **polar** mixture of water and methanol or acetonitrile. At pH 3.5 both benzoic acid and sorbic acid will be protonated (i.e. uncharged and non-polar) so interact more with the non-polar stationary

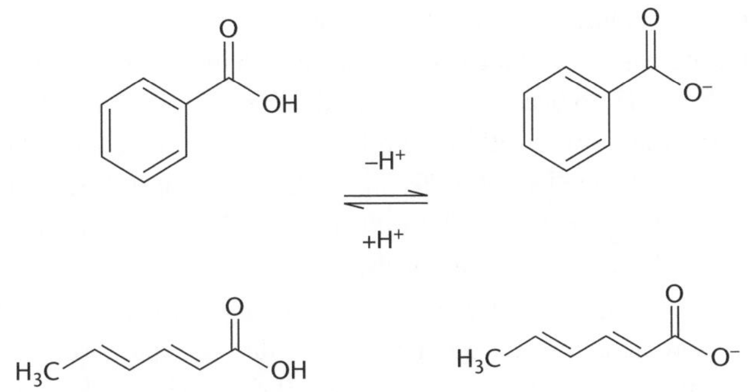

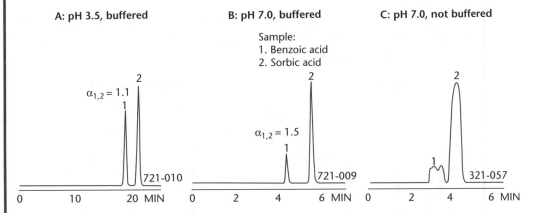

Figure 9.6 Effect of pH on the structure of ionizable compounds

A: pH 3.5, buffered   **B: pH 7.0, buffered**   **C: pH 7.0, not buffered**

Sample:
1. Benzoic acid
2. Sorbic acid

$\alpha_{1,2} = 1.1$

$\alpha_{1,2} = 1.5$

721-010   721-009   321-057

0   10   20 MIN   0   2   4   6 MIN   0   2   4   6 MIN

Figure 9.7 Effect of pH on the separation of ionizable compounds (Source: Heyrman, A.N. and Henry, R.A. 2006. *Importance of Controlling Mobile Phase pH in Reversed Phase HPLC.* Keystone Technical Bulletin TB 99-06. Bellefonte, PA: Keystone Scientific, Inc)

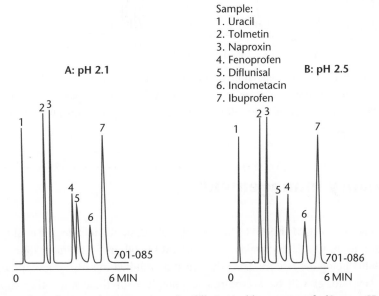

Sample:
1. Uracil
2. Tolmetin
3. Naproxin
4. Fenoprofen
5. Diflunisal
6. Indometacin
7. Ibuprofen

A: pH 2.1   **B: pH 2.5**

701-085   701-086

0   6 MIN   0   6 MIN

Figure 9.8 Effect of pH on the separation of mildly ionizable compounds (Source: Heyrman, A.N. and Henry, R.A. 2006. *Importance of Controlling Mobile Phase pH in Reversed Phase HPLC.* Keystone Technical Bulletin TB 99-06. Bellefonte, PA: Keystone Scientific, Inc)

phase. This results in a long retention time on the column of 20 min or so (Figure 9.7a). When the pH is increased to 7.0 the equilibrium shown in Figure 9.6 shifts to the right and both compounds lose H^+ to become charged and polar. This results in less interaction with the stationary phase and more with the mobile phase, so they elute quicker (Figure 9.7b). When the pH is not controlled by buffering there is less control over the charge on the compounds so the equilibrium will be influenced by local concentration effects in the column and the peaks broaden (Figure 9.7b).

Mildly ionizable compounds are not affected to such a great extent but changes in pH can have unpredictable consequences. The order of separation of fenoprofen and diflunisal shown in Figure 9.8 is reversed by a small change in pH from 2.1 to 2.5 because of the net effect on retention caused by the interactions between the non-polar part of the compounds with the stationary phase and the pH-affected polar functional groups with the mobile phase.

The retention times of non-ionizable compounds are unaffected by pH (Figure 9.9). However, other compounds in the sample may not be, so adjusting the pH can help to eliminate co-eluting impurities.

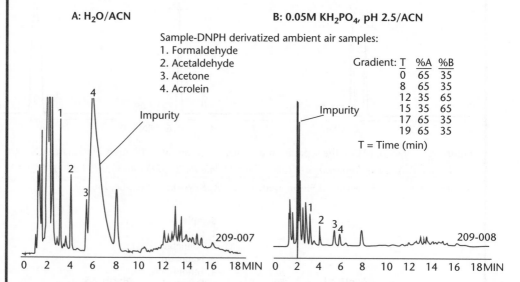

Figure 9.9 Effect of pH on separation of non-ionizable compounds (DNPH derivatized, gradient elution) (Source: Heyrman, A.N. and Henry, R.A. 2006. *Importance of Controlling Mobile Phase pH in Reversed Phase HPLC*. Keystone Technical Bulletin TB 99-06. Bellefonte, PA: Keystone Scientific, Inc)

9.3 Quality management

Quality management (QM) is a collection of activities designed to meet the quality objectives of an organization. QM is encompassed by the **ISO 9000** group of international standards, with **ISO 9001** being the requirements that must be fulfilled to satisfy these standards. Most analytical laboratories will be **accredited** to comply with these and other more specific standards, some of which are listed below:

• ISO 15189: Medical Laboratories

- ISO 17025: Measurement Laboratories (for calibration laboratories) in conjunction with ISO 15195
- ISO 17043: Proficiency Testing Scheme Providers

The elements of QM include, but are not limited to:

- Quality policy
- Quality objectives
- Quality manual
- Organizational structure and responsibilities
- Data management
- Processes
- Resources (including human)
- Product quality and customer satisfaction
- Continuous improvement
- Maintenance
- Transparency and audit

A complete discussion of these elements is beyond the scope of this book, so we shall simplify it and discuss those aspects which are directly relevant to the working analyst by adopting the **quality objective** of obtaining accurate and precise analytical results. Clearly, this is the very least that we should aim for! In order to do this we require some form of planned and systematic control which provides confidence in the analysis and data generated, known as **quality assurance**. The procedures used to provide evidence of this are called **quality control**.

9.3.1 Quality assurance

All QM systems require a great deal of **documentation**, at the heart of which is the **quality manual**. This details all of the processes which are defined by the quality management system (QMS); however, we shall only focus on particular parts of direct relevance.

9.3.1.1 Standard operating procedure

As methods are developed and validated it is necessary to document them so that other analysts can follow the procedure and obtain similar results. Also, as methods are adapted to the requirements of a particular analysis it is necessary to update the procedure and document the changes. These documentary records of analytical procedures are called **standard operating procedures (SOPs)**, and are closely related to the process of **method validation**. In this way, if analyst A in lab X performs an analysis using a specific SOP, that analyst can be confident of obtaining the same results as analyst B in lab Y.

There are many examples of standard methods for sampling, sample preparation, measurement, and data evaluation. Peer-reviewed analytical journals are a particularly important source. Where a legislative and regulatory framework is involved, **official methods** have to be followed. Some examples of sources are shown in Table 9.5.

Table 9.5 Some sources of official analytical methods

Official Source	Link to source
ISO	https://www.iso.org/
OECD	http://www.oecd.org/
EPA	https://www.epa.gov/
EA	https://www.gov.uk/government/organisations/environment-agency
DEFRA	https://www.gov.uk/government/organisations/department-for-environment-food-rural-affairs
FSA	https://www.food.gov.uk/
AOAC	https://www.aoac.org
EC Directives	https://europa.eu/european-union/index_en

ACTIVITY 9.6 Valid nitrate measurement methodology

Using our example of a need to measure the nitrate concentration in drinking water, interrogate the internet or your institution's web-based service in order to identify three recognized, valid methods that will allow you to perform this practical measurement. You will have to use suitable keywords for this search-based activity and your results must demonstrate the chemical and instrumental basis for its determination.

When you have completed this task, compare your results with your tutor's and the feedback at the end of this chapter.

Clearly, it is impractical to have an SOP for every sample type and procedure, so SOPs are often used in combination. For example, in the analysis of soil samples for poly-aromatic hydrocarbons, you could use an SOP for extraction of PAHs from soil extracts, then another for the analysis of extracts by GC-MS.

9.3.1.2 Laboratory records

You may have wondered why your tutors are keen for you to make comprehensive notes in a hard-backed laboratory notebook, rather than scribbling the results of your experiments on the back of the lab script. In a working analytical laboratory it is necessary to record every operation of a procedure so that poor results can be traced and mistakes rectified. This is part of the process called **traceability**, which was encountered in Chapter 6 and involved the use of CRMs. Often, laboratory books must be 'signed off' by the lab manager at the end of each day.

9.3.1.3 Sample register

As soon as a sample enters the lab it should be given a unique identification code so that its progress throughout the analytical process can be tracked and the chain of custody from one analyst to another can be traced. This is another part of the traceability chain; clearly you would not want a sample which is being tested for a potentially fatal disease to be mixed up with another from a healthy individual.

9.3.1.4 Test sheet

When submitting a sample for analysis the client needs to specify their requirements in sufficient detail to allow the analyst to perform the necessary test; it is not enough to request a 'full analysis' of a sample because such a thing does not exist. Test sheets can be more or less specific depending on the nature of the laboratory and the equipment available. For example, a water testing laboratory may be set up specifically to perform *Escherichia coli* determinations in drinking water, for which a test sheet will be fairly minimal, while another laboratory may offer a range of tests for many types of sample, or might be geared up to investigate suspected pollution events where a certain amount of judgement is required to narrow down the potential number of analytes before selecting a suitable procedure.

9.3.1.5 Archive

When the analysis is completed, all documentation relating to the sample, including the report to the client, must be archived for a specified period of time. This can often include the samples themselves. For example, it is often necessary to keep samples from pharmacological trials for five years or more after the analysis has been completed in case any problems arise with the product.

9.3.1.6 Calibration schedule

The calibration schedule is of utmost importance as part of the traceability chain. As well as the instrument used to perform the analysis, other pieces of laboratory equipment such as balances and volumetric pipettes must be calibrated using the appropriate CRMs. The schedule documents how frequently this must be done and the results of the calibration itself.

9.3.2 Quality control

Quality control is closely related to method validation because it involves the routine monitoring of accuracy and precision. The use of spiked samples and CRMs, which resemble the samples in matrix composition and analyte concentration, is intrinsic to this process.

9.3.2.1 Spiked samples

Spiked samples can be used to determine the **recoveries** of analytes close to the levels expected in the sample. An aliquot of the sample matrix to be analysed should be spiked with the analytes at approximately twice the naturally occurring level. Separate unspiked aliquots should also be analysed and net recoveries determined. This indicates whether the sample matrix is contributing to any bias in the results.

Spiked recoveries can be determined using the following equation:

$$R = \frac{C_s - C}{s} \times 100 \tag{9.6}$$

where R = percent recovery, C_s = spiked sample concentration, C = unspiked sample concentration, and s = concentration equivalent of analyte added to sample.

A **quality control sample** (QCS) should be analysed alongside the other samples. The QCS should contain the analytes to be determined at known concentrations and with known confidence limits and have a matrix similar to the samples to be analysed. It should be prepared in exactly the same way as the samples, and at least one should be analysed with every batch, or every 10 samples. A CRM with a similar matrix to the sample is the best QCS (see Chapter 6 for a discussion of reference materials). These will have been certified for their analyte concentration and have very well defined confidence limits. Often, it is not possible to obtain a

CRM which has been certified for every analyte required at a suitable concentration compared to the samples, and they are expensive, so an in-house reference material may be substituted instead providing that the method has been appropriately validated.

The QCS is used in conjunction with a **quality control chart** to monitor the performance of the method on a day-to-day basis to check whether it is **under control**. There are several types of control chart in common use.

9.3.2.2 Shewhart chart

There are many different ways of constructing a quality control chart. We will describe one method suitable for monitoring results of the analysis of a QCS using the same statistical method that is applied in the commercial statistical software package MINITAB®.

The chart is constructed as follows, and shown in Figure 9.10:

- If a sufficient number of **in-control measurements** of the QCS have been obtained, the **target value** can be set by calculating the mean of these results ($\bar{x}$).
- Next, it is necessary to compute the **moving range** (MR) values, which are the differences between each measurement and the previous one.

$$MR = \left| x_i - x_{i-1} \right| \tag{9.7}$$

- The average moving range ($\overline{MR}$) is then used to set **control limits** as follows:

Upper action limit (UAL) $= \bar{x} + 3\dfrac{\overline{MR}}{1.128}$ red horizontal line

Upper warning limit (UWL) $= \bar{x} + 2\dfrac{\overline{MR}}{1.128}$ amber horizontal line

Target value $= \bar{x}$ green horizontal line

Lower warning limit (LWL) $= \bar{x} - 2\dfrac{\overline{MR}}{1.128}$ amber horizontal line

Lower action limit (LAL) $= \bar{x} - 3\dfrac{\overline{MR}}{1.128}$ red horizontal line

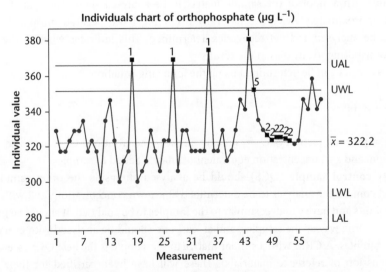

Figure 9.10 Shewhart individuals chart showing failures of tests 1, 2, and 5 (NB: the control limits are not shown in colour here, but would normally be green for the target value at 322.2, amber for +/− 2SL and red for +/− 3SL)

This type of chart is called a **Shewhart individuals chart**. The warning and action limits represent $\pm 2\sigma$ and $\pm 3\sigma$, respectively, so you would expect 3 in 1000 measurements to fall outside the action limits and 1 in 20 to fall between the warning and action limits. A problem is normally indicated when:

1. one point falls outside the action limits;
2. eight successive points fall on the same side of the mean;
3. six points in a row trend up or down;
4. 14 points in a row alternate up and down;
5. two out of the last three successive points fall outside the warning limits on the same side of the mean.

As you can see from Figure 9.10, a number of measurements have failed tests 1, 2, and 5. This alerts the analyst to potential problems in the method. In the case of failure 1, the sample may have been contaminated, a matrix or spectroscopic interference could have given an erroneously high result, or the instrument could have been incorrectly calibrated. In the cases of failures 2 and 5, contamination or an interference may be causing a slightly high result. Ideally, the results should be randomly distributed either side of the target value within the warning limits.

9.3.2.3 Moving range chart

Another way to represent the data is as a **moving range chart**. The **target value** is now $\overline{MR}$ and the control limits are set as follows, and shown in Figure 9.11:

Upper control limit	$= 3.267\,\overline{MR}$	red horizontal line
Target value	$= \overline{MR}$	green horizontal line
Lower control limit	$= 0$	red horizontal line

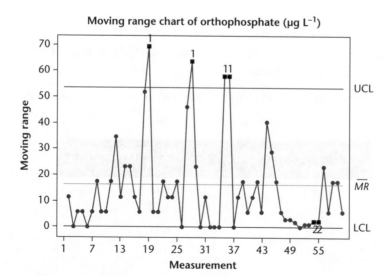

Figure 9.11 Moving range chart showing failures of tests 1 and 2. (NB: the control limits are not shown in colour here, but would normally be green for the $\overline{MR}$ target value at 16.40 and red for the LCL and UCL)

A problem is normally indicated when:

1. one point falls outside the control limits;
2. eight successive points fall on the same side of the mean;
3. six points in a row trend up or down;
4. 14 points in a row alternate up and down.

ACTIVITY 9.7 Plotting and using a quality control chart

You are an analyst working for a water company and one of your jobs is using a method to determine the amount of orthophosphate in the effluent from one of the wastewater treatment plants. Part of the monitoring procedure to check that the method is performing satisfactorily is to analyse a quality control sample with every batch of samples. The analytical data obtained from analysis of these quality control samples indicates that there are occasional problems with the method. You have been asked to investigate the problem and identify the causes.

Orthophosphate (PO_4^{3-}) reacts with molybdate in an acid medium to form a phosphomolybdate complex. This complex is then reduced by ascorbic acid, resulting in an intense blue colour (known as molybdenum blue), the absorbance of which is measured spectrophotometrically.

As part of the analytical procedure, when a batch of samples is analysed, a quality control sample, containing 6.0 mg L^{-1} PO$_4^{3-}$, is also analysed to check that the method is performing correctly. The method involves the following steps:

- accurate transfer of 10.0 mL of the sample and standard into separate 30 mL polypropylene screw-capped containers;
- addition of the contents of a reagent sachet and mixing for 15 s using a vortex mixer;
- allowing to stand for 3 min to allow the colour to fully develop;
- measuring the absorbance of the colour-developed sample and standard at 825 nm using a visible spectrophotometer.

Over a period of time a great deal of quality control data are generated. Some of these data are listed below:

Day	Orthophosphate (µg L^{-1})	Day	Orthophosphate (µg L^{-1})
1	6.05	31	6.11
2	5.56	32	6.13
3	6.02	33	6.11
4	6.49	34	6.09

Day	Orthophosphate ($\mu g \ L^{-1}$)	Day	Orthophosphate ($\mu g \ L^{-1}$)
5	6.58	35	6.20
6	6.24	36	5.51
7	6.93	37	6.18
8	6.78	38	6.09
9	5.58	39	5.58
10	5.56	40	5.96
11	6.07	41	5.80
12	5.55	42	6.27
13	6.38	43	5.56
14	5.98	44	5.96
15	6.36	45	6.05
16	6.62	46	6.38
17	6.36	47	5.51
18	5.53	48	6.29
19	5.47	49	6.93
20	5.65	50	6.95
21	5.64	51	6.20
22	5.56	52	6.58
23	5.69	53	6.38
24	6.38	54	6.51
25	7.27	55	6.22
26	6.93	56	5.53
27	6.78	57	6.47
28	6.20	58	5.27
29	6.16	59	6.13
30	6.09	60	6.87

(a) Plot a Shewhart individuals chart.

- The first 20 points have been plotted for you.

- For convenience, calculate the mean of the first 20 results and use this as the target value $\bar{x}$.

- Draw a green horizontal line on the graph (the first 20 results have been plotted for you) to represent the target value, that is, the value you would expect to be close to if the method is performing satisfactorily.

- Once again, using the first 20 results, calculate the moving range values, which are the differences between each measurement and the previous one, $MR = \left| x_i - x_{i-1} \right|$.

- Calculate the average moving range, $\overline{MR}$.

→

- Calculate the standard deviation, $\sigma = \dfrac{\overline{MR}}{1.128}$.
- Calculate the upper and lower warning limits and draw amber horizontal lines. $= \overline{x} \pm 2\dfrac{\overline{MR}}{1.128}$
- Calculate the upper and lower action limits and draw red horizontal lines $= \overline{x} \pm 3\dfrac{\overline{MR}}{1.128}$

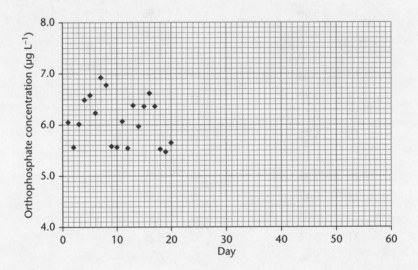

(b) Plot the rest of the data for Days 21–60 and attempt to reconcile the following with your quality control chart:

   i. Checking your lab book you discover that the quality control solution which you made up and used from Day 23 to Day 28 inclusive was of the wrong concentration.

   ii. You always use a 10 mL graduated glass pipette to transfer the sample and standard but a new graduate who has started in the laboratory has suggested that better results would be obtained by weighing.

   iii. You forgot to clean the outside of the sample cuvette on Day 25.

   iv. You allowed the new graduate to take over the analysis from Day 29 to Day 35 inclusive.

   v. You started using a new batch of reagent from Day 40 onwards.

   vi. After checking the electronic record for the spectrophotometer you discover that it was set at 815 nm by accident on Day 47.

   vii. On Days 49 and 50 you had accidently knocked over your standard solution and had to quickly prepare another while the sample was in the vortex mixer.

   viii. The air conditioning in the lab broke down on Day 55 and was not repaired for a week.

PROBLEM 9.1

Determination of diazinon in runoff from a sheep dip

You are an analyst working for the Environment Agency and have been tasked with developing a method for the determination of the organophosphorus pesticide diazinon in runoff from the area surrounding a sheep dip facility. You will be basing your method on one developed by the US Environmental Protection Agency (EPA) and are required to make a summary report to your line manager on the method development and performance. The report must include the following elements:

- a summary of the method;
- a statement that explains why the method is selective for diazinon in wastewater;
- quantitative data analysis and interpretation that demonstrates the performance of the method.

The following information is provided:

1. *EPA Method 614: The Determination of Organophosphorus Pesticides in Municipal and Industrial Wastewater*, which you can download from http://www.epa.gov/sam/index.htm by accessing the page and searching for the method.

2. calibration (Table 9.6) and quality control (Tables 9.7 and 9.8) data for the determination of diazinon in wastewater by GC;

3. a target value of 0.074 µg L^{-1} with an average moving range of 0.016 µg L^{-1}.

Table 9.6 Calibration data for the determination of diazinon by gas chromatography

Concentration of diazinon/µg L^{-1}	Peak area/cps
0	38, 24, 16, 18, 30, 34, 35, 10, 21, 10
0.04	70
0.08	140
0.4	700
0.8	1400
2	2000

Table 9.7 Replicate analysis of a quality control sample

Certified diazinon concentration/µg L^{-1}	Mean measured diazinon concentration/µg L^{-1}	s	n
0.08	0.074	±0.005	5

Table 9.8 Data for the concentration of diazinon in wastewater (in μg L^{-1}) determined over a period of 26 days

Date	Concentration of QCS/μg L^{-1}
18/06/2013	0.074
19/06/2013	0.072
20/06/2013	0.069
21/06/2013	0.074
22/06/2013	0.133
23/06/2013	0.071
24/06/2013	0.077
25/06/2013	0.087
26/06/2013	0.066
27/06/2013	0.084
28/06/2013	0.064
29/06/2013	0.076
30/06/2013	0.076
01/07/2013	0.076
02/07/2013	0.077
03/07/2013	0.084
04/07/2013	0.099
05/07/2013	0.146
06/07/2013	0.110
07/07/2013	0.092
08/07/2013	0.072
09/07/2013	0.077
10/07/2013	0.075
11/07/2013	0.069
12/07/2013	0.085
13/07/2013	0.068

Feedback on activities and problems

 FEEDBACK ON ACTIVITY 9.1

Which method?

(a) Phosphorus must be determined in lubricating oils because compounds containing phosphorus are added to decrease wear. If ICP-AES is used to directly determine phosphorus in oil, the analyte is phosphorus.

(b) Orthophosphate must be determined in river water because runoff from agricultural land which has been treated with phosphate-containing fertilizers can lead to environmental problems such as eutrophication. If UV-vis spectroscopy is used with the 'molybdenum blue' method, the analyte is orthophosphate.

(c) Phosphine, a toxic gas, must be determined in air for health and safety requirements in workplaces where it is used as a precursor in the synthesis of organophosphorus catalysts. A **sorbent tube** containing silica gel coated with mercuric cyanide is first used to trap the phosphine from air, which is then extracted and oxidized with a potassium permanganate/sulfuric acid solution and then determined as for orthophosphate using the 'molybdenum blue' method. In this case the analyte is phosphine.

(d) Diazinon, an organophosphate pesticide, must be determined in waters and soil because it can be an environmental pollutant. The analyte must be extracted from the sample and determined using gas chromatography coupled with mass spectrometry (GC-MS).

(e) Adenosine triphosphate (ATP) is the energy source for enzyme-catalysed reactions in cell metabolism. If it is determined using a luciferase **enzyme assay**, a bioluminescent reaction which generates light, the analyte is ATP.

(f) Sodium hexametaphosphate, a preservative used in foods, must be determined for quality control purposes. If liquid chromatography with a **conductivity detector** is used the analyte is hexametaphosphate.

 FEEDBACK ON ACTIVITY 9.2

Calculating the limit of detection

- *Calculate the sample standard deviation, s_0.*

    $s_0 = 0.0049 \text{ mg L}^{-1}$

- *Calculate the LOD*

    $$x_{LOD} = \frac{3s_0}{b} = \frac{3(0.0049)}{0.21} = 0.07 \text{ mg L}^{-1}$$

- *Calculate the LOQ*

    $$x_{LOQ} = \frac{10s_0}{b} = \frac{10(0.0049)}{0.21} = 0.2 \text{ mg L}^{-1}$$

Note that the LOD and LOQ are given to **one significant figure only**, which reflects the precision inherent in the measurement of the standard deviation at low concentration.

FEEDBACK ON ACTIVITY 9.3

Estimating bias using a CRM

For Analyst A:

$$t = \frac{\bar{x} - \mu}{s/\sqrt{n}}$$

$$\bar{x} = 3.89 \text{ mg kg}^{-1}$$

$$\mu = 4.14 \text{ mg kg}^{-1}$$

$$s = 1.29 \text{ mg kg}^{-1}$$

$$n = 5$$

$$\text{d.f.} = n - 1 = 4$$

$$t_{calc} = \frac{3.89 - 4.14}{1.29/\sqrt{5}} = -0.433$$

Because this is a two-sided test we want to know whether the two methods differ in any direction so we can ignore the sign of t and take the modulus:

$$\left| t_{calc} \right| = 0.433$$

The null hypothesis is that there is no difference between the measured mean and the certified value:

$$\text{H}_0: \bar{x} = \mu, \text{ which we can reject if } \left| t_{calc} \right| \geq t_{crit}$$

The critical value of t, t_{crit}, must be found from statistical tables and will depend on the level of confidence and the number of degrees of freedom:

Level of confidence, $p = 0.05$

Degrees of freedom, $\text{d.f.} = n - 1 = 4$

Hence: $t_{crit} = 2.776$

$t_{calc} < t_{crit}$, so the null hypothesis can be retained and we can say that the mean for Analyst A is not significantly different from the certified value. However, it should be noted that the precision of Analyst A's measurements is extremely poor!

For Analyst B:

$$t_{calc} = \frac{4.03 - 4.14}{0.31/\sqrt{5}} = -0.793$$

$$\left| t_{calc} \right| = 0.793$$

$$t_{crit} = 2.776$$

$t_{calc} < t_{crit}$, so the null hypothesis can be retained and we can say that the mean for Analyst B is not significantly different from the certified value.

For Analyst C:

$$t_{calc} = \frac{5.04 - 4.14}{0.29/\sqrt{5}} = 6.94$$

$$\left|t_{calc}\right| = 6.94$$

$$t_{crit} = 2.776$$

$t_{calc} > t_{crit}$, so the null hypothesis can be rejected and we can say that the mean for Analyst C is significantly different from the certified value.

? FEEDBACK ON ACTIVITY 9.4

Estimating repeatability and reproducibility

Repeatability

- *Calculate the mean, standard deviation, and relative standard deviation.*

$\bar{x} = 464.4$ mg kg^{-1}

$s = 12.4$ mg kg^{-1}

RSD = 2.7%

Reproducibility

- *Calculate the median, interquartile range, and upper and lower whiskers of a box plot. Identify any possible outliers.*

Order the data:

Lab	Mean result/μg g^{-1}
8	386
10	401
9	411
7	417
6	437
5	449
3	451
1	452
2	465
4	491

Calculate the parameters:

Median = 443

Q1 = 408.5

Q3 = 455.25

IQR = 455.25 − 408.5 = 46.75

$Q1 − (1.5 \times IQR) = 408.5 − (1.5 \times 46.75) = 408.5 − 70.125 = 338.375$

Lower whisker = 338.375 or lowest datum within $1.5 \times IQR = 386$

$Q3 + (1.5 \times IQR) = 455.25 + (1.5 \times 46.75) = 455.25 + 70.125 = 525.375$

Upper whisker = 525.375 or highest datum within $1.5 \times IQR = 491$

Draw the box and whisker plot:

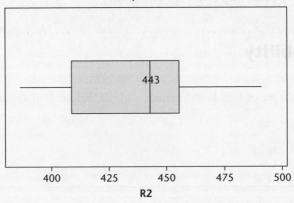

Box plot of R2

FEEDBACK ON ACTIVITY 9.5

Selectivity

(a) Phosphorus must be determined in lubricating oils because compounds containing phosphorus are added to decrease wear. If ICP-AES is used to directly determine phosphorus in oil, the method is selective for phosphorus because ICP-AES is selective for atomic emission of the element phosphorus.

(b) Orthophosphate must be determined in river water because runoff from agricultural land which has been treated with phosphate containing fertilizers can lead to environmental problems such as eutrophication. If UV-vis spectroscopy is used with the 'molybdenum blue' method, the analyte is orthophosphate. This is because the method, in which orthophosphate reacts with molybdate in an acid medium to form a phosphomolybdate complex which is then reduced by ascorbic acid resulting in an intense blue colour, is selective for orthophosphate.

(c) Phosphine, a toxic gas, must be determined in air for health and safety requirements in workplaces where it is used as a precursor in the synthesis of organophosphorus catalysts. A **sorbent tube** containing silica gel coated with mercuric cyanide is first used to trap the phosphine from air, which is then extracted and oxidized with a potassium permanganate/sulfuric acid solution and then determined as for orthophosphate using the 'molybdenum blue' method. In this case the analyte is phosphine because the method is selective for the volatile phosphine at the trapping stage.

(d) Diazinon, an organophosphate pesticide, must be determined in waters and soil because it can be an environmental pollutant. The analyte must be extracted from the sample and determined using gas chromatography coupled with mass spectrometry (GC-MS); if this method is used the analyte is diazinon because GC-MS is extremely selective for organic compounds.

(e) Adenosine triphosphate (ATP) is the energy source for enzyme catalysed reactions in cell metabolism. If it is determined using a luciferase **enzyme assay**, a bioluminescent reaction which generates light, the analyte is ATP because the assay is indirectly selective when the other substrates are controlled.

(f) Sodium hexametaphosphate, a preservative used in foods, must be determined for quality control purposes. If liquid chromatography with a **conductivity detector** is used the analyte is hexametaphosphate because selectivity is ensured by the chromatographic separation.

FEEDBACK ON ACTIVITY 9.6

Valid nitrate measurement methodology

Using our example of a need to measure the nitrate concentration in drinking water, interrogate the internet or your institution's web-based service in order to identify three recognized, valid methods that will allow you to perform this practical measurement. You will have to use suitable keywords for this search-based activity and your results must demonstrate the chemical and instrumental basis for its determination.

Three or more recognized techniques can be identified using keywords such as 'nitrate', 'determination', 'measurement', 'water', etc. The techniques used by the environmental protection agencies around the globe can include:

- ion selective electrode (ISE);
- UV-vis spectroscopy (direct method or by selective colorimetric agents);
- ion chromatography (IC) with a suitable detector.

FEEDBACK ON ACTIVITY 9.7

Plotting and using a quality control chart

(a) Plot a Shewhart individuals chart.

- *Calculate the mean of the first 20 results and use this as the target value.*

For the first 20 results:

$$\bar{x} = 6.088$$

- *Draw a green horizontal line to represent the target value, that is, the value you would expect to be close to if the method is performing satisfactorily.*

Target value = 6.088 μg L^{-1}

- *Calculate the upper and lower warning limits and draw amber horizontal lines.*

It is necessary to compute the moving range (*MR*) values for the first 20 data points and take the average.

$MR = \left| x_i - x_{i-1} \right|$ for each of the first 20 data points:

Day	Orthophosphate (μg L^{-1})	Moving range
1	6.05	
2	5.56	0.49
3	6.02	0.46
4	6.49	0.47
5	6.58	0.09
6	6.24	0.34
7	6.93	0.69
8	6.78	0.15
9	5.58	1.20
10	5.56	0.02
11	6.07	0.51
12	5.55	0.52
13	6.38	0.83
14	5.98	0.40
15	6.36	0.38
16	6.62	0.26
17	6.36	0.26
18	5.53	0.83
19	5.47	0.06
20	5.65	0.18

Now, calculate the average moving range ($\overline{MR}$):

$$\overline{MR} = 0.428$$

- *Calculate the upper and lower warning limits and draw amber horizontal lines.*
- *Calculate the upper and lower action limits and draw red horizontal lines.*

Upper action limit $= \bar{x} + 3\dfrac{\overline{MR}}{1.128} = 7.23 \ \mu g \ L^{-1}$ red horizontal line

Upper warning limit $= \bar{x} + 2\dfrac{\overline{MR}}{1.128} = 6.85 \ \mu g \ L^{-1}$ amber horizontal line

Target value $= \bar{x} = 6.088 \ \mu g \ L^{-1}$ green horizontal line

Lower warning limit $= \bar{x} - 2\dfrac{\overline{MR}}{1.128} = 5.33 \ \mu g \ L^{-1}$ amber horizontal line

Lower action limit $= \bar{x} - 3\dfrac{\overline{MR}}{1.128} = 4.95 \ \mu g \ L^{-1}$ red horizontal line

Your Shewhart chart should look like the one below [(NB: the control limits are not shown in colour here, but would normally be green for the target value at 6.09, amber for the upper and lower warning limits (UWL and LWL), and red for the upper and lower action limits (UAL and LAL)]:

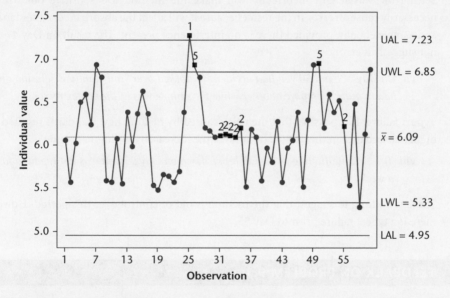

(b) Plot the rest of the data for Days 21–60 and attempt to reconcile the following with your quality control chart:

i. Checking your lab book you discover that the quality control solution which you made up and used from Day 23 to Day 28 inclusive was of the wrong concentration.

Type 1 and 5 failure. It is possible that that the quality control solution is incorrect but you should be able to check your lab book to see whether this will give a high or low result. However, there is a trend up and down rather than a consistent error so there may be another contributing factor.

ii. You always use a 10 mL graduated glass pipette to transfer the sample and standard but a new graduate who has started in the laboratory has suggested that better results would be obtained by weighing.

A grade-A graduated pipette will be accurate to half a division, or 0.05 mL so your measurement volume is accurate to at least 3 significant figures. The results are reported to three significant figures so it is unlikely that the precision of the results will be improved by using an analytical balance, provided that the pipette has been calibrated and used correctly.

iii. You forgot to clean the outside of the sample cuvette on Day 25.

Type 1 failure. A dirty sample cuvette will cause a higher absorbance reading and hence a high result for the sample, which corresponds with the result outside the action limit on Day 25.

iv. You allowed the new graduate to take over the analysis from Day 29 to Day 35 inclusive.

Type 2 failure. However, the results between Day 29 and Day 35 are consistently close to the target value—give the graduate a pay rise.

v. You started using a new batch of reagent from Day 40 onwards.

There is nothing to suggest that a new batch of reagent has caused the method to go out of control.

vi. After checking the electronic record for the spectrophotometer you discover that it was set at 815 nm by accident on Day 47.

Setting the wavelength incorrectly will make the method less sensitive but will not necessarily cause an error in the result because it will affect the absorbance of the standard and sample equally, provided there is no interference present. The result on Day 47 does not suggest an error.

vii. On Days 49 and 50 you had accidently knocked over your standard solution and had to quickly prepare another while the sample was in the vortex mixer.

Type 5 failure. It is possible that the delay caused by making new standards resulted in a higher absorbance in the sample because of the increased development time.

viii. The air conditioning in the lab broke down on Day 55 and was not repaired for a week.

There is nothing to suggest that the method is out of control over this period. However, there is a type 2 failure prior to Day 55.

 FEEDBACK ON PROBLEM 9.1

Determination of diazinon in runoff from a sheep dip

You are an analyst working for the Environment Agency and have been tasked with developing a method for the determination of the organophosphorus pesticide diazinon in runoff from the area surrounding a sheep dip facility. You will be basing your method on one developed by the US Environmental Protection Agency (EPA) and are required to make a summary report to your line manager on the method development and performance. The report must include the following elements:

- *A summary of the method*

Your summary should include the following key points:

- – transfer into separation funnel;
- – extraction three times into the organic phase;
- – drying;
- – transfer into the Kuderna-Danish (K-D) concentrator equipped with a 10 mL concentrator tube, a 500 mL evaporation flask, and Snyder column;
- – method of operation of the K-D concentrator to end up with 1–2 mL in hexane.

All steps in that order together with volumes and timings should be given.

- *A statement that explains why the method is selective for diazinon in wastewater*

The method is selective for diazinon for three reasons: the extraction procedure effects some separation of the non-polar compounds from the aqueous phase; the gas chromatography instrument separates these extracted compounds from each other prior to detection; and a phosphorus-specific detector (FPD) is used.

- *Quantitative data analysis and interpretation that demonstrates the performance of the method*

You should use the data to evaluate the following performance characteristics:

Linear dynamic range

This is between the LOD (or LOQ) and the upper linear part of the calibration curve, which should be plotted to show this.

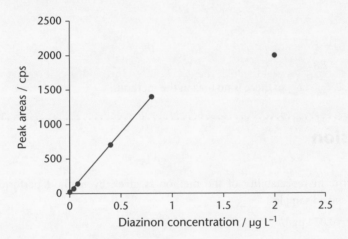

The calibration is only linear up to 0.8 μg L^{-1}, so the slope of the linear part of the curve should be used.

Limit of detection

Slope, b	=	1736 cps μg^{-1} L		
Intercept, a	=	8.30 cps		
s	=	10.29		
$3s$	=	30.9		
y_0 (mean)	=	23.6		
y_{LOD}	=	$y_0 + 3s$	=	54.5
LOD	=	$3s/b$	=	0.02 μg L^{-1}

Limit of quantitation

LOQ $=$ $10s/b$ $=$ $0.06\ \mu g\ L^{-1}$

Bias

Using the method performance data for diazinon spiked samples, a t-test can be performed.

$$t = \frac{\bar{x} - \mu}{s/\sqrt{n}}$$

$\bar{x} = 0.074\ \mu g\ L^{-1}$

$\mu = 0.08\ \mu g\ L^{-1}$

$s = 0.005$

$n = 5$

d.f. $= 4$

$t_{calc} = -2.7$

$|t_{calc}| = 2.7$

$t_{crit} = 2.8$

In this case, $t_{calc} < t_{crit}$ so there is no bias in the method.

Precision

The short-term repeatability of the method is given by method performance data for diazinon spiked samples.

$\bar{x} = 0.074\ \mu g\ L^{-1}$

$\mu = 0.08\ \mu g\ L^{-1}$

$s = 0.005$

$n = 5$

RSD $= 0.005/0.074 \times 100\% = 7\%$

The long-term repeatability of the method is given by the quality control data.

$\bar{x} = 0.083$

$s = 0.020$

$n = 26$

RSD $= 0.020/0.083 \times 100\% = 24\%$

Ruggedness

No data show this.

Performance over time

A Shewhart individuals chart should be drawn to show performance of the method over time. This should be plotted with a target value of 0.074 µg L^{-1}, and upper and lower warning and action limits calculated using the average moving range divided by 1.128.

$\overline{MR} = 0.0163$

Upper action limit = 0.117 µg L^{-1}

Upper warning limit = 0.103 µg L^{-1}

Target value = 0.074 µg L^{-1}

Lower warning limit = 0.045 µg L^{-1}

Lower action limit = 0.031 µg L^{-1}

Your Shewhart chart should look like the one below.

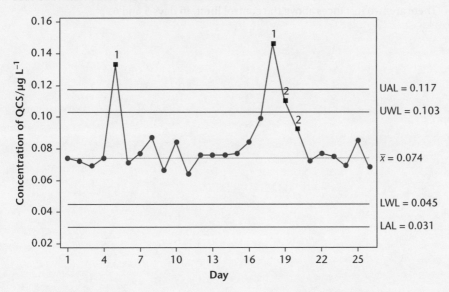

- There are two instances above the action limits, in days 5 and 18.

- There are eight successive points on the same side of the mean around the high value at day 18, which indicates that there may have been a problem with the method around this period, particularly given the gradual increase and decrease in the concentration.

- The precision of the QA data is similar to the precision of the target value.

Alternatively, a moving range chart can be drawn. This should be plotted with an average moving range target value and upper and lower confidence limits calculated using the average moving range multiplied by 3.267.

Upper control limit $= 3.267\,\overline{MR} = 0.052$ µg L^{-1}

Target value $= \overline{MR} = 0.0163$ µg L^{-1}

Lower control limit $= 0$ µg L^{-1}

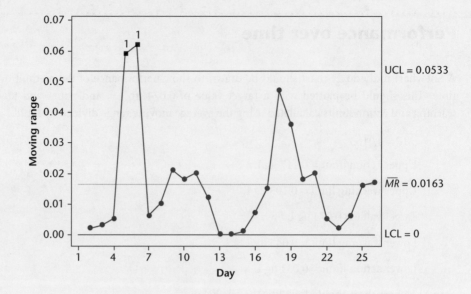

- There are two instances above the control limit, in days 5 and 6.

Appendix A

Critical values for *t*- and *F*-tests

Table A1 Critical values of *t*

d.f. One sided Two sided	0.90 0.80	0.95 0.90	0.975 0.95	0.995 0.99	0.9975 0.995	0.9995 0.999
1	3.078	6.314	12.706	63.657	127.321	639.619
2	1.886	2.920	4.303	9.925	14.089	31.598
3	1.638	2.353	3.182	5.841	7.453	12.941
4	1.533	2.132	2.776	4.604	5.598	8.610
5	1.476	2.015	2.571	4.032	4.773	6.859
6	1.440	1.943	2.447	3.707	4.317	5.959
7	1.415	1.895	2.365	3.499	4.029	5.405
8	1.397	1.860	2.306	3.355	3.833	5.041
9	1.384	1.833	2.262	3.250	3.690	4.781
10	1.372	1.812	2.228	3.169	3.581	4.587
11	1.363	1.796	2.201	3.106	3.497	4.437
12	1.356	1.782	2.179	3.055	3.428	4.318
13	1.350	1.771	2.160	3.012	3.372	4.221
14	1.345	1.761	2.145	2.977	3.326	4.140
15	1.341	1.753	2.131	2.947	3.286	4.073
16	1.337	1.746	2.120	2.921	3.252	4.015
17	1.333	1.740	2.110	2.898	3.222	3.965
18	1.330	1.734	2.101	2.878	3.197	3.922
19	1.328	1.729	2.093	2.861	3.174	3.883
20	1.325	1.725	2.086	2.845	3.153	3.850
21	1.323	1.721	2.080	2.831	3.135	3.819
22	1.321	1.717	2.074	2.819	3.119	3.792
23	1.320	1.714	2.069	2.807	3.104	3.767
24	1.318	1.711	2.064	2.797	3.091	3.745
25	1.316	1.708	2.060	2.787	3.078	3.725

d.f. One sided Two sided	0.90 0.80	0.95 0.90	0.975 0.95	0.995 0.99	0.9975 0.995	0.9995 0.999
26	1.315	1.706	2.056	2.779	3.067	3.707
27	1.314	1.703	2.052	2.771	3.057	3.690
28	1.313	1.701	2.048	2.763	3.047	3.674
29	1.311	1.699	2.045	2.756	3.038	3.659
30	1.310	1.697	2.042	2.750	3.030	3.646
40	1.303	1.684	2.021	2.705	2.971	3.551
50	1.299	1.676	2.009	2.678	2.937	3.496
60	1.296	1.671	2.000	2.660	2.915	3.460
70	1.294	1.667	1.994	2.648	2.899	3.435
80	1.292	1.664	1.990	2.639	2.887	3.416
90	1.291	1.662	1.987	2.632	2.878	3.402
100	1.290	1.660	1.984	2.626	2.871	3.391
120	1.289	1.658	1.980	2.617	2.860	3.373
150	1.287	1.655	1.976	2.609	2.849	3.357
200	1.286	1.653	1.972	2.601	2.839	3.340
∞	1.282	1.645	1.960	2.576	2.807	3.291

Table A2 Critical values of F (one-sided): $p = 0.95$ (upper); $p = 0.975$ (middle); $p = 0.99$ (lower)

d.f.	1	2	3	4	5	6	7	8	9	10	11	12	13	14	15	16	17	18	19	20
1	161	200	216	225	230	234	237	239	241	242	243	244	245	245	246	246	247	247	248	248
	648	799	864	900	922	937	948	957	963	969	973	977	980	983	985	987	989	990	992	993
	4052	5000	5403	5625	5764	5859	5928	5981	6022	6056	6083	6106	6126	6143	6157	6170	6181	6192	6201	6209
2	18.5	19.0	19.2	19.2	19.3	19.3	19.4	19.4	19.4	19.4	19.4	19.4	19.4	19.4	19.4	19.4	19.4	19.4	19.4	19.4
	38.5	39.0	39.2	39.2	39.3	39.3	39.4	39.4	39.4	39.4	39.4	39.4	39.4	39.4	39.4	39.4	39.4	39.4	39.4	39.4
	98.5	99.0	99.2	99.2	99.3	99.3	99.4	99.4	99.4	99.4	99.4	99.4	99.4	99.4	99.4	99.4	99.4	99.4	99.4	99.4
3	10.13	9.55	9.28	9.12	9.01	8.94	8.89	8.85	8.81	8.79	8.76	8.74	8.73	8.71	8.70	8.69	8.68	8.67	8.67	8.66
	17.4	16.0	15.4	15.1	14.9	14.7	14.6	14.5	14.5	14.4	14.4	14.3	14.3	14.3	14.3	14.2	14.2	14.2	14.2	14.2
	34.1	30.8	29.5	28.7	28.2	27.9	27.7	27.5	27.3	27.2	27.1	27.1	27.0	26.9	26.9	26.8	26.8	26.8	26.7	26.7
4	7.71	6.94	6.59	6.39	6.26	6.16	6.09	6.04	6.00	5.96	5.94	5.91	5.89	5.87	5.86	5.84	5.83	5.82	5.81	5.80
	12.2	10.6	10.0	9.6	9.36	9.20	9.07	8.98	8.90	8.84	8.79	8.75	8.71	8.68	8.66	8.63	8.61	8.59	8.58	8.56
	21.2	18.0	16.7	16.0	15.5	15.2	15.0	14.8	14.7	14.5	14.5	14.4	14.3	14.2	14.2	14.2	14.1	14.1	14.0	14.0
5	6.61	5.79	5.41	5.19	5.05	4.95	4.88	4.82	4.77	4.74	4.70	4.68	4.66	4.64	4.62	4.60	4.59	4.58	4.57	4.56
	10.0	8.43	7.76	7.39	7.15	6.98	6.85	6.76	6.68	6.62	6.57	6.52	6.49	6.46	6.43	6.40	6.38	6.36	6.34	6.33
	16.3	13.3	12.1	11.4	11.0	10.7	10.5	10.3	10.2	10.1	10.0	9.9	9.8	9.8	9.7	9.7	9.6	9.6	9.6	9.6
6	5.99	5.14	4.76	4.53	4.39	4.28	4.21	4.15	4.10	4.06	4.03	4.00	3.98	3.96	3.94	3.92	3.91	3.90	3.88	3.87
	8.81	7.26	6.60	6.23	5.99	5.82	5.70	5.60	5.52	5.46	5.41	5.37	5.33	5.30	5.27	5.24	5.22	5.20	5.18	5.17
	13.7	10.9	9.78	9.15	8.75	8.47	8.26	8.10	7.98	7.87	7.79	7.72	7.66	7.60	7.56	7.52	7.48	7.45	7.42	7.40
7	5.59	4.74	4.35	4.12	3.97	3.87	3.79	3.73	3.68	3.64	3.60	3.57	3.55	3.53	3.51	3.49	3.48	3.47	3.46	3.44
	8.07	6.54	5.89	5.52	5.29	5.12	4.99	4.90	4.82	4.76	4.71	4.67	4.63	4.60	4.57	4.54	4.52	4.50	4.48	4.47
	12.2	9.55	8.45	7.85	7.46	7.19	6.99	6.84	6.72	6.62	6.54	6.47	6.41	6.36	6.31	6.28	6.24	6.21	6.18	6.16

d.f.	1	2	3	4	5	6	7	8	9	10	11	12	13	14	15	16	17	18	19	20
8	5.32	4.46	4.07	3.84	3.69	3.58	3.50	3.44	3.39	3.35	3.31	3.28	3.26	3.24	3.22	3.20	3.19	3.17	3.16	3.15
	7.57	6.06	5.42	5.05	4.82	4.65	4.53	4.43	4.36	4.30	4.24	4.20	4.16	4.13	4.10	4.08	4.05	4.03	4.02	4.00
	11.3	8.65	7.59	7.01	6.63	6.37	6.18	6.03	5.91	5.81	5.73	5.67	5.61	5.56	5.52	5.48	5.44	5.41	5.38	5.36
9	5.12	4.26	3.86	3.63	3.48	3.37	3.29	3.23	3.18	3.14	3.10	3.07	3.05	3.03	3.01	2.99	2.97	2.96	2.95	2.94
	7.21	5.71	5.08	4.72	4.48	4.32	4.20	4.10	4.03	3.96	3.91	3.87	3.83	3.80	3.77	3.74	3.72	3.70	3.68	3.67
	10.6	8.02	6.99	6.42	6.06	5.80	5.61	5.47	5.35	5.26	5.18	5.11	5.05	5.01	4.96	4.92	4.89	4.86	4.83	4.81
10	4.96	4.10	3.71	3.48	3.33	3.22	3.14	3.07	3.02	2.98	2.94	2.91	2.89	2.86	2.85	2.83	2.81	2.80	2.79	2.77
	6.94	5.46	4.83	4.47	4.24	4.07	3.95	3.85	3.78	3.72	3.66	3.62	3.58	3.55	3.52	3.50	3.47	3.45	3.44	3.42
	10.0	7.56	6.55	5.99	5.64	5.39	5.20	5.06	4.94	4.85	4.77	4.71	4.65	4.60	4.56	4.52	4.49	4.46	4.43	4.41
11	4.84	3.98	3.59	3.36	3.20	3.09	3.01	2.95	2.90	2.85	2.82	2.79	2.76	2.74	2.72	2.70	2.69	2.67	2.66	2.65
	6.72	5.26	4.63	4.28	4.04	3.88	3.76	3.66	3.59	3.53	3.47	3.43	3.39	3.36	3.33	3.30	3.28	3.26	3.24	3.23
	9.65	7.21	6.22	5.67	5.32	5.07	4.89	4.74	4.63	4.54	4.46	4.40	4.34	4.29	4.25	4.21	4.18	4.15	4.12	4.10
12	4.75	3.89	3.49	3.26	3.11	3.00	2.91	2.85	2.80	2.75	2.72	2.69	2.66	2.64	2.62	2.60	2.58	2.57	2.56	2.54
	6.55	5.10	4.47	4.12	3.89	3.73	3.61	3.51	3.44	3.37	3.32	3.28	3.24	3.21	3.18	3.15	3.13	3.11	3.09	3.07
	9.33	6.93	5.95	5.41	5.06	4.82	4.64	4.50	4.39	4.30	4.22	4.16	4.10	4.05	4.01	3.97	3.94	3.91	3.88	3.86
13	4.67	3.81	3.41	3.18	3.03	2.92	2.83	2.77	2.71	2.67	2.63	2.60	2.58	2.55	2.53	2.51	2.50	2.48	2.47	2.46
	6.41	4.97	4.35	4.00	3.77	3.60	3.48	3.39	3.31	3.25	3.20	3.15	3.12	3.08	3.05	3.03	3.00	2.98	2.96	2.95
	9.07	6.70	5.74	5.21	4.86	4.62	4.44	4.30	4.19	4.10	4.02	3.96	3.91	3.86	3.82	3.78	3.75	3.72	3.69	3.66
14	4.60	3.74	3.34	3.11	2.96	2.85	2.76	2.70	2.65	2.60	2.57	2.53	2.51	2.48	2.46	2.44	2.43	2.41	2.40	2.39
	6.30	4.86	4.24	3.89	3.66	3.50	3.38	3.29	3.21	3.15	3.09	3.05	3.01	2.98	2.95	2.92	2.90	2.88	2.86	2.84
	8.86	6.51	5.56	5.04	4.69	4.46	4.28	4.14	4.03	3.94	3.86	3.80	3.75	3.70	3.66	3.62	3.59	3.56	3.53	3.51

d.f.	1	2	3	4	5	6	7	8	9	10	11	12	13	14	15	16	17	18	19	20
15	4.54	3.68	3.29	3.06	2.90	2.79	2.71	2.64	2.59	2.54	2.51	2.48	2.45	2.42	2.40	2.38	2.37	2.35	2.34	2.33
	6.20	4.77	4.15	3.80	3.58	3.41	3.29	3.20	3.12	3.06	3.01	2.96	2.92	2.89	2.86	2.84	2.81	2.79	2.77	2.76
	8.68	6.36	5.42	4.89	4.56	4.32	4.14	4.00	3.89	3.80	3.73	3.67	3.61	3.56	3.52	3.49	3.45	3.42	3.40	3.37
16	4.49	3.63	3.24	3.01	2.85	2.74	2.66	2.59	2.54	2.49	2.46	2.42	2.40	2.37	2.35	2.33	2.32	2.30	2.29	2.28
	6.12	4.69	4.08	3.73	3.50	3.34	3.22	3.12	3.05	2.99	2.93	2.89	2.85	2.82	2.79	2.76	2.74	2.72	2.70	2.68
	8.53	6.23	5.29	4.77	4.44	4.20	4.03	3.89	3.78	3.69	3.62	3.55	3.50	3.45	3.41	3.37	3.34	3.31	3.28	3.26
17	4.45	3.59	3.20	2.96	2.81	2.70	2.61	2.55	2.49	2.45	2.41	2.38	2.35	2.33	2.31	2.29	2.27	2.26	2.24	2.23
	6.04	4.62	4.01	3.66	3.44	3.28	3.16	3.06	2.98	2.92	2.87	2.82	2.79	2.75	2.72	2.70	2.67	2.65	2.63	2.62
	8.40	6.11	5.18	4.67	4.34	4.10	3.93	3.79	3.68	3.59	3.52	3.46	3.40	3.35	3.31	3.27	3.24	3.21	3.19	3.16
18	4.41	3.55	3.16	2.93	2.77	2.66	2.58	2.51	2.46	2.41	2.37	2.34	2.31	2.29	2.27	2.25	2.23	2.22	2.20	2.19
	5.98	4.56	3.95	3.61	3.38	3.22	3.10	3.01	2.93	2.87	2.81	2.77	2.73	2.70	2.67	2.64	2.62	2.60	2.58	2.56
	8.29	6.01	5.09	4.58	4.25	4.01	3.84	3.71	3.60	3.51	3.43	3.37	3.32	3.27	3.23	3.19	3.16	3.13	3.10	3.08
19	4.38	3.52	3.13	2.90	2.74	2.63	2.54	2.48	2.42	2.38	2.34	2.31	2.28	2.26	2.23	2.21	2.20	2.18	2.17	2.16
	5.92	4.51	3.90	3.56	3.33	3.17	3.05	2.96	2.88	2.82	2.76	2.72	2.68	2.65	2.62	2.59	2.57	2.55	2.53	2.51
	8.18	5.93	5.01	4.50	4.17	3.94	3.77	3.63	3.52	3.43	3.36	3.30	3.24	3.19	3.15	3.12	3.08	3.05	3.03	3.00
20	4.35	3.49	3.10	2.87	2.71	2.60	2.51	2.45	2.39	2.35	2.31	2.28	2.25	2.22	2.20	2.18	2.17	2.15	2.14	2.12
	5.87	4.46	3.86	3.51	3.29	3.13	3.01	2.91	2.84	2.77	2.72	2.68	2.64	2.60	2.57	2.55	2.52	2.50	2.48	2.46
	8.10	5.85	4.94	4.43	4.10	3.87	3.70	3.56	3.46	3.37	3.29	3.23	3.18	3.13	3.09	3.05	3.02	2.99	2.96	2.94

Table A3. Critical values of F (two-sided): $p = 0.95$ (upper); $p = 0.975$ (middle); $p = 0.99$ (lower)

d.f.	1	2	3	4	5	6	7	8	9	10	11	12	13	14	15	16	17	18	19	20
1	648	799	864	900	922	937	948	957	963	969	973	977	980	983	985	987	989	990	992	993
	2593	3200	3458	3600	3689	3750	3794	3828	3854	3876	3893	3908	3920	3931	3941	3949	3956	3962	3968	3973
	16211	20000	21615	22500	23056	23437	23715	23925	24091	24224	24334	24426	24505	24572	24630	24681	24727	24767	24803	24836
2	38.5	39.0	39.2	39.2	39.3	39.3	39.4	39.4	39.4	39.4	39.4	39.4	39.4	39.4	39.4	39.4	39.4	39.4	39.4	39.4
	78.5	79.0	79.2	79.2	79.3	79.3	79.4	79.4	79.4	79.4	79.4	79.4	79.4	79.4	79.4	79.4	79.4	79.4	79.4	79.4
	199	199	199	199	199	199	199	199	199	199	199	199	199	199	199	199	199	199	199	199
3	17.4	16.0	15.4	15.1	14.9	14.7	14.6	14.5	14.5	14.4	14.4	14.3	14.3	14.3	14.3	14.2	14.2	14.2	14.2	14.2
	29.1	26.3	25.2	24.6	24.2	23.9	23.7	23.6	23.5	23.4	23.3	23.2	23.1	23.1	23.1	23.0	23.0	23.0	22.9	22.9
	55.6	49.8	47.5	46.2	45.4	44.8	44.4	44.1	43.9	43.7	43.5	43.4	43.3	43.2	43.1	43.0	42.9	42.9	42.8	42.8
4	12.2	10.6	10.0	9.6	9.36	9.20	9.07	8.98	8.90	8.84	8.79	8.75	8.71	8.68	8.66	8.63	8.61	8.59	8.58	8.56
	18.6	15.9	14.8	14.1	13.8	13.5	13.3	13.1	13.0	12.9	12.8	12.8	12.7	12.7	12.6	12.6	12.5	12.5	12.5	12.5
	31.3	26.3	24.3	23.2	22.5	22.0	21.6	21.4	21.1	21.0	20.8	20.7	20.6	20.5	20.4	20.4	20.3	20.3	20.2	20.2
5	10.0	8.4	7.8	7.4	7.15	6.98	6.85	6.76	6.68	6.62	6.57	6.52	6.49	6.46	6.43	6.40	6.38	6.36	6.34	6.33
	14.5	11.9	10.9	10.3	9.90	9.64	9.45	9.31	9.19	9.10	9.02	8.95	8.90	8.85	8.81	8.77	8.74	8.71	8.68	8.66
	22.8	18.3	16.5	15.6	14.9	14.5	14.2	14.0	13.8	13.6	13.5	13.4	13.3	13.2	13.1	13.1	13.0	13.0	12.9	12.9
6	8.81	7.26	6.60	6.23	5.99	5.82	5.70	5.60	5.52	5.46	5.41	5.37	5.33	5.30	5.27	5.24	5.22	5.20	5.18	5.17
	12.4	9.9	8.9	8.4	8.0	7.75	7.56	7.42	7.31	7.22	7.14	7.08	7.02	6.98	6.94	6.90	6.87	6.84	6.81	6.79
	18.6	14.5	12.9	12.0	11.5	11.1	10.8	10.6	10.4	10.3	10.1	10.0	10.0	9.88	9.81	9.76	9.71	9.66	9.62	9.59
7	8.1	6.5	5.9	5.5	5.29	5.12	4.99	4.90	4.82	4.76	4.71	4.67	4.63	4.60	4.57	4.54	4.52	4.50	4.48	4.47
	11.1	8.7	7.8	7.2	6.88	6.64	6.46	6.32	6.21	6.12	6.05	5.99	5.93	5.89	5.85	5.81	5.78	5.75	5.73	5.70
	16.2	12.4	10.9	10.1	9.52	9.16	8.89	8.68	8.51	8.38	8.27	8.18	8.10	8.03	7.97	7.91	7.87	7.83	7.79	7.75

d.f.	1	2	3	4	5	6	7	8	9	10	11	12	13	14	15	16	17	18	19	20
8	7.57	6.06	5.42	5.05	4.82	4.65	4.53	4.43	4.36	4.30	4.24	4.20	4.16	4.13	4.10	4.08	4.05	4.03	4.02	4.00
	10.28	7.96	7.02	6.49	6.15	5.92	5.74	5.61	5.50	5.41	5.34	5.28	5.23	5.18	5.14	5.11	5.08	5.05	5.02	5.00
	14.69	11.04	9.60	8.81	8.30	7.95	7.69	7.50	7.34	7.21	7.10	7.01	6.94	6.87	6.81	6.76	6.72	6.68	6.64	6.61
9	7.21	5.71	5.08	4.72	4.48	4.32	4.20	4.10	4.03	3.96	3.91	3.87	3.83	3.80	3.77	3.74	3.72	3.70	3.68	3.67
	9.68	7.42	6.49	5.98	5.65	5.41	5.24	5.11	5.00	4.92	4.85	4.79	4.73	4.69	4.65	4.62	4.58	4.56	4.53	4.51
	13.61	10.11	8.72	7.96	7.47	7.13	6.88	6.69	6.54	6.42	6.31	6.23	6.15	6.09	6.03	5.98	5.94	5.90	5.86	5.83
10	6.94	5.46	4.83	4.47	4.24	4.07	3.95	3.85	3.78	3.72	3.66	3.62	3.58	3.55	3.52	3.50	3.47	3.45	3.44	3.42
	9.23	7.01	6.10	5.60	5.27	5.04	4.88	4.74	4.64	4.56	4.48	4.42	4.37	4.33	4.29	4.25	4.22	4.20	4.17	4.15
	12.83	9.43	8.08	7.34	6.87	6.54	6.30	6.12	5.97	5.85	5.75	5.66	5.59	5.53	5.47	5.42	5.38	5.34	5.31	5.27
11	6.72	5.26	4.63	4.28	4.04	3.88	3.76	3.66	3.59	3.53	3.47	3.43	3.39	3.36	3.33	3.30	3.28	3.26	3.24	3.23
	8.89	6.70	5.81	5.31	4.99	4.76	4.60	4.47	4.36	4.28	4.21	4.15	4.10	4.05	4.01	3.98	3.95	3.92	3.90	3.87
	12.23	8.91	7.60	6.88	6.42	6.10	5.86	5.68	5.54	5.42	5.32	5.24	5.16	5.10	5.05	5.00	4.96	4.92	4.89	4.86
12	6.55	5.10	4.47	4.12	3.89	3.73	3.61	3.51	3.44	3.37	3.32	3.28	3.24	3.21	3.18	3.15	3.13	3.11	3.09	3.07
	8.61	6.45	5.57	5.08	4.76	4.54	4.37	4.25	4.14	4.06	3.99	3.93	3.88	3.83	3.80	3.76	3.73	3.70	3.68	3.66
	11.75	8.51	7.23	6.52	6.07	5.76	5.52	5.35	5.20	5.09	4.99	4.91	4.84	4.77	4.72	4.67	4.63	4.59	4.56	4.53
13	6.41	4.97	4.35	4.00	3.77	3.60	3.48	3.39	3.31	3.25	3.20	3.15	3.12	3.08	3.05	3.03	3.00	2.98	2.96	2.95
	8.39	6.26	5.38	4.90	4.58	4.36	4.20	4.07	3.97	3.88	3.81	3.75	3.70	3.66	3.62	3.59	3.56	3.53	3.50	3.48
	11.37	8.19	6.93	6.23	5.79	5.48	5.25	5.08	4.94	4.82	4.72	4.64	4.57	4.51	4.46	4.41	4.37	4.33	4.30	4.27
14	6.30	4.86	4.24	3.89	3.66	3.50	3.38	3.29	3.21	3.15	3.09	3.05	3.01	2.98	2.95	2.92	2.90	2.88	2.86	2.84
	8.20	6.09	5.23	4.74	4.43	4.21	4.05	3.92	3.82	3.74	3.67	3.61	3.56	3.51	3.48	3.44	3.41	3.38	3.36	3.34
	11.06	7.92	6.68	6.00	5.56	5.26	5.03	4.86	4.72	4.60	4.51	4.43	4.36	4.30	4.25	4.20	4.16	4.12	4.09	4.06

d.f.	1	2	3	4	5	6	7	8	9	10	11	12	13	14	15	16	17	18	19	20
15	6.20	4.77	4.15	3.80	3.58	3.41	3.29	3.20	3.12	3.06	3.01	2.96	2.92	2.89	2.86	2.84	2.81	2.79	2.77	2.76
	8.05	5.95	5.10	4.62	4.31	4.09	3.93	3.80	3.70	3.62	3.55	3.49	3.44	3.39	3.36	3.32	3.29	3.26	3.24	3.22
	10.80	7.70	6.48	5.80	5.37	5.07	4.85	4.67	4.54	4.42	4.33	4.25	4.18	4.12	4.07	4.02	3.98	3.95	3.91	3.88
16	6.12	4.69	4.08	3.73	3.50	3.34	3.22	3.12	3.05	2.99	2.93	2.89	2.85	2.82	2.79	2.76	2.74	2.72	2.70	2.68
	7.91	5.83	4.98	4.51	4.20	3.98	3.82	3.70	3.60	3.51	3.44	3.39	3.33	3.29	3.25	3.22	3.19	3.16	3.14	3.11
	10.58	7.51	6.30	5.64	5.21	4.91	4.69	4.52	4.38	4.27	4.18	4.10	4.03	3.97	3.92	3.87	3.83	3.80	3.76	3.73
17	6.04	4.62	4.01	3.66	3.44	3.28	3.16	3.06	2.98	2.92	2.87	2.82	2.79	2.75	2.72	2.70	2.67	2.65	2.63	2.62
	7.80	5.73	4.89	4.42	4.11	3.89	3.73	3.61	3.51	3.42	3.36	3.30	3.25	3.20	3.16	3.13	3.10	3.07	3.05	3.02
	10.38	7.35	6.16	5.50	5.07	4.78	4.56	4.39	4.25	4.14	4.05	3.97	3.90	3.84	3.79	3.75	3.71	3.67	3.64	3.61
18	5.98	4.56	3.95	3.61	3.38	3.22	3.10	3.01	2.93	2.87	2.81	2.77	2.73	2.70	2.67	2.64	2.62	2.60	2.58	2.56
	7.70	5.65	4.80	4.33	4.03	3.82	3.65	3.53	3.43	3.35	3.28	3.22	3.17	3.13	3.09	3.05	3.02	2.99	2.97	2.95
	10.22	7.21	6.03	5.37	4.96	4.66	4.44	4.28	4.14	4.03	3.94	3.86	3.79	3.73	3.68	3.64	3.60	3.56	3.53	3.50
19	5.92	4.51	3.90	3.56	3.33	3.17	3.05	2.96	2.88	2.82	2.76	2.72	2.68	2.65	2.62	2.59	2.57	2.55	2.53	2.51
	7.61	5.57	4.73	4.26	3.96	3.75	3.59	3.46	3.36	3.28	3.21	3.15	3.10	3.06	3.02	2.98	2.95	2.93	2.90	2.88
	10.07	7.09	5.92	5.27	4.85	4.56	4.34	4.18	4.04	3.93	3.84	3.76	3.70	3.64	3.59	3.54	3.50	3.46	3.43	3.40
20	5.87	4.46	3.86	3.51	3.29	3.13	3.01	2.91	2.84	2.77	2.72	2.68	2.64	2.60	2.57	2.55	2.52	2.50	2.48	2.46
	7.53	5.50	4.67	4.20	3.90	3.69	3.53	3.40	3.30	3.22	3.15	3.09	3.04	3.00	2.96	2.93	2.89	2.87	2.84	2.82
	9.94	6.99	5.82	5.17	4.76	4.47	4.26	4.09	3.96	3.85	3.76	3.68	3.61	3.55	3.50	3.46	3.42	3.38	3.35	3.32

Glossary

accuracy Closeness of the agreement between the result of a measurement and a true (or accepted) value of the measurand (where measurand is the particular quantity subject to measurement).

amount of substance In the chemical sense refers to the number of primary or elementary entities (e.g. atoms, molecules, ions, particles etc.) divided by the Avogadro constant.

analyte The entity (e.g. atom, molecule, ion, functional group, etc.) being investigated. This can be both qualitatively and quantitatively.

analytical approach A framework that follows a logical series of steps in order to solve (answer) an analytical problem (question).

bias The difference between the measured and true value.

bulk or 'bulk sample' is a portion/an amount of material taken from a larger source. The source may be the primary material or site or a sub-division from which further sub-divisions are made. The term 'bulk' or 'bulked' to mean 'composited increments' is to be avoided.

calibration The measurement of standard solutions or materials in order to construct a calibration curve.

calibration curve A graphical representation of response plotted as a function of some property of a standard, e.g. concentration or absolute amount of analyte.

calibration function The equation of the line of the calibration curve.

certified reference material (CRM) A material characterized by a metrologically valid procedure for one or more specified properties, accompanied by a certificate that provides the value of the specified property, its associated uncertainty, and a statement of metrological traceability.

chromatography A method of separating mixtures whereby the components to be separated are physically distributed between a stationary phase (e.g. a solid or immobile liquid) and a flowing mobile phase (e.g. liquid or gas).

chromophores Sections of molecules which can undergo detectable electronic transitions in the 'UV to visible' part of the spectrum; often, a group of atoms as a functional group or a bond-type are responsible for this effect.

coefficient of determination, R^2 The square of the correlation coefficient.

composite sample A sample, ideally made up from equal quantities of different samples or increments, but of the same source material.

contamination The introduction of undesirable material during the analytical process which may or may not be the analyte of interest.

correlation coefficient, r A measure of the strength and direction of a linear relationship between two variables.

degrees of freedom (d.f.) The number of degrees of freedom used to estimate a statistical parameter is the number of independent measurements that go into the estimate.

detector A device which responds to the presence of a particular substance or analyte, usually arranged to produce an electronic signal.

digestion The process of breaking down a sample in order to change its physical state and release the analyte of interest for further processing or measurement.

dissolution The process of converting a sample into the liquid phase using a suitable solvent or chemical reactant.

extraction Separation of the analyte from a sample matrix.

fit for purpose Where an analytical method can be demonstrated to be valid for its intended purpose.

flux A substance, usually an inorganic salt, used in a fusion procedure to dissolve refractory compounds.

fusion The chemical reaction between a sample and a flux material used to dissolve refractory compounds in the molten state, usually resulting in a solid bead or pellet when cooled to room temperature.

gross sample A portion of material taken from a larger source which may be the primary material.

heterogeneous sample A sample that demonstrates variation in its composition or properties throughout its extent.

homogeneous sample A sample that does not demonstrate variation in its composition or properties throughout its extent.

interference A systematic error caused by interferents in the sample or other extraneous sources.

interferent A component, usually in the sample but can be extraneous, which affects the measurement.

ISO9001 Part of the ISO9000 series of standards for quality management.

isobaric isotopes of different elements that have the same nominal mass

judgemental sample A sample chosen based upon prior knowledge or judgement in order to evaluate a particular property.

limit of detection, LOD The lowest amount of analyte which can be detected, but not necessarily quantified.

limit of quantitation, LOQ The lowest amount of analyte which can be quantified in a sample with reasonable confidence.

linear range The range between the lower and upper concentration of analyte in the sample for which the method is applicable; often taken to mean the range between the LOD, or the LOQ, and the upper linear point on the calibration curve.

lot or batch Usually indicates a known or definite quantity of a produced or manufactured material for sampling.

matrix Everything in the sample except the analyte.

matrix interference An interference caused by the matrix component(s).

matrix matching A procedure whereby calibration standards are prepared in a matrix of similar composition to the sample.

method validation The process of determining the performance characteristics of a method to

establish whether the analytical results obtained will be fit for their intended purpose.

performance characteristic A characteristic of the method to be determined during method validation, e.g. LOD.

precision The measure of the spread of results obtained under specified conditions, made by making repeated independent measurements on identical samples.

pre-concentration A process in which the sample is treated in such a way as to increase the analyte concentration.

primary sample The sample taken from the bulk material.

procedural blank A sample obtained by carrying out all of the steps of the analytical procedure in the absence of the sample.

qualitative analysis The process of determining the identity of the sample and/or presence of the analyte.

quality assurance The process of planning and systematic control designed to provide confidence in the analytical methods and results.

quality control Procedures used to monitor performance of an analytical method in order to provide confidence in the analysis.

quality control chart A graphical record of the results of the analysis of a quality control sample.

quality control sample A sample of known composition, similar in matrix and analyte concentration to the samples to be analysed.

quantitative analysis The process of measuring the amount of the analyte that is present in the sample.

random errors Variations in a measured value that occur by chance.

random sample A sample chosen from the bulk material at random.

recovery factor The fraction (or percentage) of the analyte obtained after one or more stages of a method.

reference material A material, sufficiently homogeneous and stable with respect to one or more specified properties, which has been established to be fit for its intended use in a measurement process.

repeatability The precision of the results of independent measurements that have been performed on identical samples under the same conditions (e.g. same operator, same apparatus, same laboratory) over a short time interval.

representative sample A sample expected to adequately demonstrate the properties of interest of the parent material.

reproducibility The precision of the results of the analysis of identical samples under different conditions (different operator, different apparatus, different laboratory) over a long time interval.

sample The material to be analysed containing the analyte or property to be determined.

sample increment An individual quantity or portion of material acquired through a single operation using a selected sampling device and usually taken in a process separated in time or space.

sample integrity How close the state of the sample, and analytes therein, are to their original form.

sample preparation A process which encompasses a variety of procedures designed to turn the sample into a form suitable for analysis.

sampling The process of collecting and storing samples for analysis.

sampling error A gross error in the measurement introduced at the sampling stage.

sampling plan A formalised procedure for the collection and storage of samples that can be followed in such a way as to make it fit for purpose.

selective sample A sample chosen to deliberately demonstrate a particular aspect of the material from which it is taken or to exclude certain characteristics.

selectivity The extent to which other substances affect the determination of an analyte; closely related to specificity.

sensitivity The change in instrument signal (or response) per unit change in amount of analyte; in practical terms this is given by the slope of the calibration curve.

serial dilution A procedure whereby increasingly dilute solutions of a stock solution are prepared, usually for the purpose of preparing standard solutions.

specificity The ability to discriminate between the analyte and everything else present in the matrix.

spiked sample A sample prepared by adding a known quantity of analyte to a matrix which is close or identical with that of the sample of interest.

standard additions A procedure used to determine the amount of analyte in a sample by analysis of spiked samples.

standard solution A solution containing a known amount of analyte used for calibration; usually prepared from a stock solution of a certified reference material.

stock solution A concentrated solution of analyte, usually prepared from a certified reference material, used to make standard solutions by serial dilution.

stratified sampling Samples obtained from identified subparts (strata) of the parent population; within each stratum, the samples are taken randomly.

sub-sample Part of an initial sample taken from the source or at a later stage.

systematic error The difference between the measured value and the true value.

systematic sampling A form of probability sampling where increments from a larger population are selected using a fixed periodic interval but from a random starting point.

traceability The process whereby the result of analysis of an unknown sample can be traced to an established and agreed value, e.g. analytical balances are traceable to the SI kilogram.

true value The actual concentration of an analyte in a sample; this cannot normally be determined.

under control When a method is performing in a consistent manner it is deemed to be under statistical control.

working curve See linear range.

working range See linear range.

Index

Page numbers in italics refer to figures and tables; n indicates that the indexed term is in a footnote.